Women and War

WOMEN AND WAR

Jean Bethke Elshtain

Basic Books, Inc., Publishers

NEW YORK

Excerpts from "The First Year of My Life" from *The Stories of Muriel Spark* (New York: E.P. Dutton, 1985), pp. 263–69. It was first published in *The New Yorker*. Reprinted with permission.

Quotes from James Axtell from "The Vengeful Women of Marblehead: Robert Roules's Deposition of 1677," *William and Mary Quarterly* 31 (1974): 647–52.

Library of Congress Cataloging-in-Publication Data

Elshtain, Jean Bethke, 1941–
   Women and war.

   Bibliographical notes: p. 259.
   Includes index.
   1. Women and war.   2. Elshtain, Jean Bethke,
1941–   . I. Title.
U21.5.E45   1987   303.6'6'088042   86–47731
ISBN   0–465–09214–4

To the memory of

John Lennon

and to the grandchildren of Paul G. Bethke and Helen Lind Bethke

Sheri, Heidi, Jenny, and Eric Elshtain

John, Mark, and Paul Bing

Timothy, John, Shana, and Susan (Soo-Jin) Stegner

Bryan and Michelle Bethke

and those yet to come

CONTENTS

Contents

PART II

LIFE GIVERS/LIFE TAKERS:

HISTORY'S GENDER GAP

Contents

PREFACE

THE TITLE of this book, *Women and War*, may seem straight-forward. In fact, it is not. Neither *women* nor *war* is a self-evident category. For my purposes, war is an object of discourse central to historic understandings of politics in the West. Without war stories there would be many fewer stories to tell. War is a structure of experience; a form of conflict; a pervasive presence. I am not concerned with origins and standard histories in taking on war as a theme; rather, I am involved—as a political theorist, a citizen, and a mother—in what *we* continue to make of war. How do we treat the war stories deeded to us? What representations of war remain resonant and resilient?

That takes me to women, not as the female members of the human race but as a complex construction. By speaking of women *and* war—rather than presenting a chronology of women *in* wars or a paean to the notion that women have not been in the thick of violent things, not eagerly anyhow—I signal my intention to explore diverse discourses and the political claims and social identities they sustain. Women have played many parts in narratives of war and politics. In the stories, theories, and events I shall discuss, women are represented as beings laced through and through with sexual and maternal imagery, including the residues of an everlasting and often intimate combat and cooperation with men—in war *and* peace. In the sense I here evoke, women have structured conflicts and collaborations, have crystallized and imploded what successive epochs imagine when the subject at hand is collective violence.

The representations I interrogate are ones we are all heir to, drawing them in as children in our families and schools, absorbing them through movies, newspapers, television, and texts. Contesting the terrain that identifies and gives meaning to our received understandings of women, men, and war does not grant a self-subsisting autonomy to discourse; rather, it implies a recognition of the ways in which received war stories may lull our critical faculties to sleep. For finally there is the terrible fact that while war is everywhere deplored, what war promises is both openly and secretly coveted.

Five years ago, before I began this book, I had a much clearer notion of what constituted folly and what wisdom, what madness and what sanity, where matters of war and peace and male and female involvement in organized, purposeful violence were concerned. I considered military women, for example, a misguided lot for placing their bodies at the behest of the state. But then I remembered my own childhood hankering for wartime testing and began to appreciate what some women seek in military life, though I doubt most find it. I was compelled to acknowledge diversity among military women; they are no more of a piece than are civilian women. Similarly, I found briefly compelling the formulation that military men are little more than a gang of grown-up boys with deadly toys. Although some men no doubt conform to the image of adolescents itching for a fight, many others seek ways to constrain and limit violence and find nuclear weapons repulsive. *Women and War* is the result of overlapping recognitions of the complexity hiding behind many of our simple, rigid ideas and formulations.

Unfortunately, contemporary social science is often ill equipped to understand the constitutive role of symbols, myths, metaphors, and rhetorical strategies, preferring instead the apparent solidity of institutional arrangements, the regularity of codified rules, or the reassurance of abstract models. Much that is important and subtle falls through the grid of standard modes and methods and is ignored. Interpretive daring is precluded. My method, if it can be called that, is not unlike Hannah Arendt's description of her own approach. She charmingly dubbed it *Perlenfischerei*, "pearl fishing." One dives in, she said, not knowing quite what one will come up with. The important point is to remain open to one's subject matter, to see where it is going and follow—not to impose a prefabricated formula over diverse and paradoxical material.

Preface

I begin, in an introductory chapter, with a brief look at the way our dominant symbols of male fighters and female noncombatants, recognizable as the Just Warrior and the Beautiful Soul, continue to resonate in contemporary discourse and in popular understanding. The seductions of these prototypical male and female constructions is such that they figure in both feminist and anti-feminist argument. In chapter 1, I play on, and with, the classic war story as a first-person narration—the heroism, tragedy, or pathos of one who has been there. Otherwise, how could he presume to tell others what it was really like in the trenches, on the beaches, in country?* My story is not-a-soldier's story; it is, instead, a tale of how war, rumors of war, and images of violence, individual and collective, permeated the thoughts of a girl growing up in a particular place—Timnath, Colorado, population 185—in the 1950s and coming of age in a very different place, near Boston, Massachusetts, in the 1960s. The narrative in this chapter is purposely episodic, fractured: first, because it was the only way for me to remain faithful to my material and to the self that takes shape in and through the text; and, second, because I wanted to present an alternative to the teleological structure of traditional war stories with their foreordained movement from brave and fearful beginnings to triumphant or tragic endings.

The chapters that comprise part I involve a complex tracking of the shifting construction of war as sanctioned collective violence, a discourse of *armed* civic virtue, and the ways in which that discourse got embodied in historic persons and events. The ideas I examine are pervasive features of political realities, shaping the self-understandings and actions of men and women in diverse times and places. Chapter 2 moves back and forth between past and present, traversing theoretical and polemical terrain, showing how the dominant terms and tropes of politics as war, war as politics, figure even among those who oppose war and often among those who propose change. In chapter 3, I explore exemplary tales of armed civic virtue in Civil War America and the era of the First World War. An exemplary tale as I use it—and there are several scattered throughout my discussion—is a concrete illustration of the dense filiations of women with war and of war as politics.

---

* "In country" is the way American soldiers referred to Vietnam.

Stories of both sorts figure in chapter 3. As I move on to chapter 4, I consider attempts to disarm civic virtue within just-war formulations and pacifist hopes. The warrior, in this scheme of things, must either throw away his weapons entirely or fight fair and square by rules that limit violence even in the midst of war. The noncombatant female, on the other hand, becomes history's Beautiful Soul, a collective being embodying values and virtues at odds with war's destructiveness, representing home and hearth and the humble verities of everyday life. I explore present implications of the just-war tradition, on the one hand, and the privatization of women's identities in the modern West, on the other.

Part II gets to the heart and body of the matter as I examine the traditional dichotomy whereby women are seen as the life givers, men as the life takers. Thus I take up, in chapter 5, women who have sent sons and husbands off to war, sometimes eagerly, seeking their honor in Spartan motherhood, and their unlikely, but nonetheless real, sisters, whom I call the Ferocious Few. Tales of the Noncombatant Many dominate our thinking, but stories of female fighters vie for attention in my discussion. Male identities, as these revolve around war fighting, are interpreted in chapter 6 as I look at the many who fight, the few who refuse to bear arms, and—perhaps most interesting of all—the Compassionate Warrior. I detect certain similarities in the structure of male war experience and female maternal experience: the Good Soldier and the Good Mother turn out to have something in common. To make my case, I look to first-person accounts; novels and poems; anthropology and myth. Why have so many men and women both loved and hated war? What is at stake in this ambivalence?

Finally, in chapter 7, I consider in detail themes that have surfaced earlier, namely, diverse modes of liberal and feminist argumentation concerning men, women, and war. Here, as elsewhere, the picture is blurred. In the matter of collective violence, as in every other, feminists are diverse and do not speak with a single voice. What is at stake in the current debates is not only the meaning and aims of feminism, but whether we might move beyond a discourse dominated by war and its necessary antithesis, peace. For peace, too, is a problem in terms of how it has been understood and what yearnings have been encircled by its promise. I challenge the Just Warrior/Beautiful Soul formulation, the symbiosis between fighter and victim presumed by the language of war and peace. Whatever the virtues and strengths of these images, we must move beyond such congealed typifications.

How can or might we relocate ourselves in order to create space for a less rigid play of individual and civic virtues than those we have known? What alternative images of citizenship can we draw upon? What perspectives currently within our reach offer hope for sustaining an ethos, stripped of utopian pretensions, that extends the prospect of limiting force and the threat of force? These are among the questions that present possibilities as we traverse the perplexing terrain of our tumultuous present.

In thinking and sometimes dreaming about war over the past few years; in reading war stories and watching war movies; in composing portions of chapters on walks as well as on my word processor, I have gained a heightened awareness of the fleeting preciousness of life, including the lives we humans share with the other creatures with whom we have yet to learn to live in decency. My next project flows from this one.

ACKNOWLEDGMENTS

MY grateful appreciation, first, to the Institute for Advanced Study, Princeton, New Jersey, for a year of uninterrupted research and contemplation, 1981–82—a fruitful interlude in which I did much of the preliminary reading and rumination for *Women and War*—and, second, to the Bellagio Conference and Study Center of the Rockefeller Foundation, under whose auspices I spent a pleasant three weeks in 1983 contemplating Lake Como from my terrace and working through alternative formulae for the architecture of this book.

I have many individuals to thank for their support, their encouragement, and, in some instances, their friendly criticism. I shall leave it up to the persons involved to sort out into which category they belong. My gratitude, then, to Michael Walzer, William Connolly, Bradley Klein, Nancy Huston, Catharine Stimpson, Sara Ruddick; to my colleagues, Jerry King, James Der Derian, and Nicholas Xenos; to faculty and students, undergraduate and graduate, some known to me, others remembered faces attached to unknown names, at several dozen universities and colleges, from Santa Cruz in the west to the University of Maine, Machias, in the far northeast, for posing questions and pressing criticisms over the past five years as this book took shape. I am also indebted to students at the University of Massachusetts, Amherst, who have signed up for my course "Issues of War and Peace in a Nuclear Age" over the past three years; I have learned much from them. No author is the sole creator of his or her work. This book is a collaborative effort. Its themes and concerns have been forged in and through public dialogue as well as private contemplation.

Offering invaluable assistance at various stages of this project were

Acknowledgments

my student assistant, Tom Concannan, who raced back and forth to various libraries for books and helped with the index; my son, Eric, who typed reference notes and checked quotations; and my husband, Errol, who saw me through copy editing and illness.

Without the determined needling of the president of Basic Books and my editor, Martin Kessler, I might still be reading yet another half-dozen works on war, or on women, or on warriors, or on all three. The supply of sources, primary and secondary, is endless, and my preparation tended toward mimesis. Mr. Kessler knew when and where to push and how to press questions that made this work far more incisive, in its own rather free-form way, than it would otherwise have been. No one could ask for a more conscientious and creative copy editor than Phoebe Hoss. I was awed by her attunement to the details as well as to the overall structure of this manuscript.

Women and War

Introduction

Beautiful Souls/Just Warriors:
The Seduction of War

W E LIVE TODAY, at the end of the twentieth century, in a world increasingly polarized, between light and dark, between "them" and "us," between women and men, with nuclear war looming as the most terrible form of potential collective destruction, but with group violence occurring all over the world, always with one justification or another. In a time of dislocation, the Manichean view—we, the "good," versus them, the "bad"—is, though comfortable, also false and dangerous. False, as I myself know, remembering a little girl who wanted a gun and a brother who did not; a father who could never kill a deer and a hunting brother who does not see the sport in war. Dangerous because this simplistic view depends on rigid notions of what men and women are in relation to war and of war itself as an absolute contrast to peace.

War seduces us in part because we continue to locate ourselves *inside* its prototypical emblems and identities. Men fight as avatars of a nation's sanctioned violence. Women work and weep and sometimes protest within the frame of discursive practices that turn one out, militant mother and pacifist protestor alike, as the collective "other" to

3

the male warrior. These identities are underpinnings for decision and action, nonetheless real for being symbolic. It is my contention that such "constellations of enshrined ideas"—to use Clifford Geertz's term for the animating descriptions of our social world, our political lives[1]—entangle us in webs of anticipated actions and reactions.

We in the West are the heirs of a tradition that assumes an affinity between women and peace, between men and war, a tradition that consists of culturally constructed and transmitted myths and memories. Thus, in time of war, real men and women—locked in a dense symbiosis, perceived as beings who have complementary needs and exemplify gender-specific virtues—take on, in cultural memory and narrative, the personas of Just Warriors and Beautiful Souls. Man construed as violent, whether eagerly and inevitably or reluctantly and tragically; woman as nonviolent, offering succor and compassion: these tropes on the social identities of men and women, past and present, do not denote what men and women *really* are in time of war, but function instead to re-create and secure women's location as noncombatants and men's as warriors. These paradigmatic linkages dangerously overshadow other voices, other stories: of pacific males; of bellicose women; of cruelty incompatible with just-war fighting; of martial fervor at odds—or so we choose to believe—with maternalism in women. No conscious bargain was struck by our collective foremothers and fathers to ensure this outcome. Rather, sedimented lore—stories of male war fighters and women home keepers and designated weepers over war's inevitable tragedies—have spilled over from one epoch to the next.

Take, for example, the female Beautiful Soul. The locution "beautiful soul" is Hegel's which I deploy here for my own purposes. In his *Phenomenology of Spirit,* Hegel characterizes the "beautiful soul" as a being defined by a mode of consciousness which allows him or her to protect "the appearance of purity by cultivating innocence about the historical course of the world."[2] Although Hegel does not make the case—having other fish to fry—I was struck upon rereading him with the recognition that women in Western culture have served as collective, culturally designated "beautiful souls." The cultivation of socially sanctioned innocence about the world's ways is not, of course, woman's own doing in any simple sense. But it is partly her doing in a complex sense: that is, the image is continually being reconstructed by women and reinforced by men.

Two samples of political rhetoric illustrate this mutually reinforcing dynamic. The first quote is drawn from a speech by a male anti-suffragist; the second, from one of the great Elizabeth Cady Stanton's rhetorical masterworks. At odds over suffrage, Stanton and her less eloquent counterpart nonetheless *share* a deep symbolic structure, each, to different ends, tapping Beautiful Soul images and imperatives.*

The nineteenth-century male anti-suffragist urges:

> Man assumed the direction of government and war, woman of the domestic and family affairs and the care and training of the child. . . . It has been so from the beginning . . . and it will continue to be so to the end, because it is in conformity to nature and its laws, and is sustained and confirmed by the experience and reason of six thousand years. . . . The domestic altar is a sacred flame where woman is the high and officiating priestess. . . . To keep her in that *condition of purity*, it is necessary that she should be separated from the exercise of suffrage and from all those stern and contaminating and demoralizing duties that devolves upon the hardier sex—man.[4] (Italics added.)

In a second representative "anti" speech, the "true woman" is referred to as "the queen of home and of hearts, . . . above the political collision of this world." One determined "anti," having lost all restraint, runs an analogy that is unintentionally hilarious. He declares the human male analogous to a bull whose bullish qualities must not be "demasculinized," diluted "with the qualities of the cow." The bull/man bears a heavy burden, as the cow/woman's "herd safety" depends on him.[5]

Countering the "antis'" opposition to suffrage, many suffragists went on to embrace the Beautiful Soul symbolism they shared *with* opponents of women's suffrage. They did so, in part, because they believed it, or some version of it; in part, because the imagery *did* capture features of what women really were or had become; in part, because they hoped they could turn it to their political advantage by predicting that the world of the warriors, if it continued on its way untamed by the virtues of Beautiful Souls, would lead humanity to

---

* There were female opponents of the Beautiful Soul. For the radical and anarchist writer and activist Emma Goldman, the formulation also meant opposition to suffrage as just another way middle-class women enslave themselves. Suffrage she proclaims a hoax, and suffragists who would purify politics as "essentially" purists, "naturally bigoted and relentless in her [the Beautiful Soul's] effort to make others as good as she thinks they ought to be."[3]

certain ruin. Women's nurturing qualities, now muted, must come out into the public light to purify politics and to tip the balance to peace and decency.

Thus Cady Stanton in 1868:

> The male element is a destructive force, stern, selfish, aggrandizing, loving war, violence, conquest, acquisition, breeding in the material and moral world alike discord, disorder, disease and death. See what a record of blood and cruelty the pages of history reveal! . . . The male element has held high carnival thus far, it has fairly run riot from the beginning, overpowering the feminine element everywhere, crushing out the diviner qualities in human nature. . . . The need of this hour is not territory, gold mines, railroads, or specie payments, but a new evangel of womanhood, to exalt purity, virtue, morality, true religion, to lift man up into higher realms of thought and action.[6]

Christopher Lasch notes that such assumptions coincided neatly with those of the "antis" and reinforced congealed male and female typologies. Although it was abstractly possible for suffragists to seek out and accumulate historic evidence suggesting that, in fact, the divide between life givers and life takers was less total than either they or the anti-suffragists claimed, neither side was prepared to challenge systematically the symbolic structures each shared. For when we speak of the constitutive role of prototypical symbols, we refer not to interests people may have, or to rational calculations of possible costs and benefits they may compute, but to what in fact people *are* and have become: it is a question of *identities*, not easily sloughed off external garments.

I write these words in 1986, nearly 120 years after Elizabeth Cady Stanton delivered her resonant, alliterative speech. We have been witness to and participant in collective violence of such catastrophic proportions that those who search for words adequate to the task of description profess failure.* If violence is the measure of an age, ours is a nadir if it is anything at all. In a post-Holocaust, post-Hiroshima world, can Beautiful Souls and Just Warriors retain their luster? Do these symbols any longer animate or tap parts of the modern self, male or female? Has the emergence of the woman as a mobilized war worker,

---

* See, for example, Kurt Vonnegut's rueful comments on his inability to capture the Dresden firestorm in novel form.[7]

a soldier, a terrorist, a revolutionary, or the prime victim of total war (along with her children) shattered our notions of Beautiful Souls or Just Warriors?

One finds conflicting evidence. For example: in her important study, *Mobilizing Women for War*, Leila J. Rupp argues that, in the Second World War, neither the Nazi mother making munitions for her son nor America's Rosie the Riveter made a permanent dent on women's economic status or on the public's basic beliefs about women's "nature" in either society, respectively. Beneath her overalls, Rosie was still "wearing her apron" in the expectation that demobilization would restore the *status quo ante*.[8] D'Ann Campbell, in *Women at War with America*, goes even further. Her reconstruction of extensive empirical studies convinces her that women of the Second World War era anticipated and helped to promulgate what became the reigning cultural ideal of postwar domesticity. During the war, housewives and war workers alike expressed essentially private dreams as their most cherished postwar desire.[9] Campbell tells a tale of complex social negotiations between and among men and women as husbands and wives, demobilized workers and soldiers. In other words, women were not crudely coerced *en masse* to "return" to the home: most of them had never left it to begin with, and those who had shared the domestic dreams of those who had not. My point for now is that the postwar romance with domesticity ensured the survival of reigning symbols, as did the fact that returning soldiers were greeted by a unified people as just and honored warriors.

We are four decades beyond the Second World War, the last "good" war (see pages 189–91). Much has changed or been challenged, including received expectations about the social identities and proper spheres for men and women. But it would be unwise to assume that the combined effects of Vietnam, feminism, the involvement of over 50 percent of adult American women in the labor force, and the growing postponement of marriage and childbirth by young women undercut received webs of social meaning as these revolve around men, women, and war.* Feminist protest has pushed several directions in

---

* While "women in the work force" statistics conjure up the image of full-time employment, the majority of women who work do so part-time, some just a few hours a week. These are,

this matter. Some feminists proclaim a "right to fight": they, too, can be Just Warriors. Others, appalled at refashioning the new woman after the old man, consciously appropriate traditional notions of women's connection to ideals of peace as a cultural basis for anti-militarist activism. To the extent that feminist pacificism reproduces Beautiful Soul presumptions and evocations it, paradoxically, helps to ensure the continued triumph of our grand narrative of women and war, of male and female identities insofar as these are forged around the matter of collective violence. The "solution," if change is one's aim, would seem to be insisting that women, too, can take up arms. But it is unclear just how far this would take us. Western history is dotted with tales of those I call the Ferocious Few, women who reversed cultural expectations by donning warrior's garb and doing battle; and their existence as fact and myth seems not to have put much of a dent in the overall edifice of the way war figures in the structure of male and female experience and reactions.

What has been badly battered is the notion that in soldiering lies duty, honor, and sacrifice. Sacrifice, yes, we are likely to respond. But duty? We as a society are not unanimous in concurring. As to honor, that has an archaic ring, trailing in its wake the slightly musty scent of an artifact too long in a dusty, cobwebbed closet.* Instead, it is the case that, when most of us look at modern war, we see collective devastation rather than individual honor. Glory, too, is in short supply. The home front can no longer be protected: the Beautiful Soul and her children are vulnerable in light of current realities. Other countries, countries marred unlike our own by the scars of war, have understood this since the Second World War. As well, John Keegan, in *The Face of Battle*, argues that the modern battlefield's inhuman face, a nightmare of destructive technological might, reduces the warrior's role "to that of mere victim." The destructive force of our weapons overshadows and overtakes our fighting men: "Impersonality, coercion, deliberate cruelty, all deployed on a rising scale, make the fitness of modern man to sustain the stress of battle increasingly doubtful."[11] The techniques

---

however, counted in the over-50-percent figure. I note this in order not to present so sweeping a vision of social change as to misrepresent social realities.

  * Peter Berger has lamented the loss of honor and the toll this loss has taken in human identity.[10]

of modern warfare have not stopped war, of course, but they have changed warrioring. The chief limiting factor in much of the technology of modern and future war has become "man himself," according to experts who describe the ways in which the F-16 jet-fighter "can dish out more than we can take, both physically and mentally."[12] The gravitational force involved in the F-16 is beyond the level of most human endurance—as likely to make its pilot suddenly unconscious as it is to serve him in a fight against the "enemy."

For Americans, the impact of what is being dubbed the "Vietnam experience," means it is now less possible in any unambiguous way to picture our war fighters as Just Warriors than it was before 1965. Whether Americans in Vietnam, Israelis in South Lebanon, Russians in Afghanistan, whether by accident or design, the modern warrior fights in a context that blurs distinctions between combatants and noncombatants, making ring more and more hollow his army's proclaimed determination to abide by rules of engagement consistent with received *jus in bello* rules, which stipulate where, how, and whom soldiers can kill in a war-fighting situation.

I wonder how deeply Americans have been scarred by past historic tragedies or take note of stark contemporary realities. Evidence points different directions. Reassurances about home-front invincibility, to which President Reagan has given such powerful voice with his SDI initiative (the "Star Wars" antiballistic-missile defense proposal), are sustained by the fact that America's warriors have always come home to a land unscarred by bombs, to intact homes, and to the whole bodies of their loved ones. Surely, we think, we can find foolproof ways to protect ourselves against Soviet missiles!

I think we Americans remain believers, abashed and embarrassed perhaps, but believers nonetheless. War retains the power to incite parts of the self that peace cannot seem to reach. Wartime's Beautiful Soul is no ordinary wife or mother or secretary or nurse: she becomes a civic being; she is needed by others; she can respond simultaneously to what Jane Addams called the "family claim" and the "social claim," for, she is told, without her unselfish devotion to country and family each would be lost. Might it be possible to crank up this vision again, to make it stick as a compelling ethos for most women? Many women yearn for public identities as civic beings serving ends that take them beyond the borders of private lives. One deeply rooted conviction, sustained by veterans of battle fronts and home fronts and transmitted to others, is that wars—good wars that unite us—offer a communal endeavor, the sharing of sacrifice and danger. Modern society appears to have found no other way to initiate and sustain action in common with others on this scale.

J. Glenn Gray recounts the story of a Frenchwoman whom he had known in war years, a time of danger and suffering, and then met again, years later, in a time of peace and comfort. She told Gray: "You know that I do not love war or want it to return. But at least it made me feel alive, as I have not felt alive before or since."[13] A nurse veteran of Vietnam—an activist organizing readjustment counseling for the nearly eight thousand women who were stationed in Vietnam, many of whom suffer from the delayed-stress syndrome widely recognized in Vietnam combat vets—acknowledges all the horror, but then tells a reporter for the *New York Times*: "I think about Vietnam often and I find myself wishing I was back there. Life over there was so real and in some ways so much easier. There was no such thing as black or white, male or female. We dealt with each other as human beings, as friends. We worked hard, we partied hard, we were a unit. A lot of us, when we left, wished we didn't have to come home."[14]

Although the seductions of war are greatest for would-be soldiers—adolescent men wondering if they will get a chance to prove themselves in "our kind of war" fought fair and square by the rules of the game—women are by no means immune to the battle call of Mars. Women who dream of being warriors perhaps long to have major roles rather than supporting parts in future war stories, in "the paradigm of all narratives." They would get up "on the stage of History.... They prefer to abandon weeping for weapons."[15] Anti-militarist feminists, working to counter seductive images of female soldiering, speak derisively of the "Private Benjamin syndrome" and describe it in devastating terms.

The syndrome gets its name from the hit film, released in 1981 (see photo section, #4). A rich, pretty, spoiled "princess," her husband the victim of a fatal heart attack on their wedding night, bumbles mistakenly into the army. She whines and moans and demands special privileges. She wants to go out to lunch. She wants to go home. But the army succeeds in making a woman out of her; indeed, the transformation is so complete that when she is about to make the mistake of marrying a second time, to a wealthy, charming but feckless Frenchman, Private Benjamin pulls herself together and stalks out of her own wedding, striding determinedly back into the waiting arms of the U.S. Army. That, at least, is the deromanticized interpretation.[16*]

I, too, found this film disturbing, a rosy, portrait of army life that seems a far cry from accounts proffered by military women themselves. Nevertheless, the attractions of the film should also be acknowledged. Benjamin is compelled, for the first time in her spoiled life, to live on terms of rough equality with others as her pleas for special privilege are denied. Also for the first time, she is placed in situations where others depend on her, and her actions make a real difference to the outcome of a group effort. When she becomes a heroine to her female buddies, the communal delight seems real enough. She discovers strengths and skills previously untapped. Perhaps my account of the film's allure owes too much to all those stories by men explaining why they loved war and combat even as they hated it. But to turn a blind eye to expressions of hate *and* love for war, wartime, and army life voiced by combatants and noncombatants alike means that, by definition, one will fall short in one's understanding of "why war?" To acknowledge the space that war offers, or promises, for action in common with others, even to the extent of forfeiting one's life to spare one's fellows, makes politics more complicated even as it enhances the range of that understanding.

A second box-office hit, released in 1982, refurbishes our historic romance with Warriors and Beautiful Souls. *An Officer and a Gentleman* repeats the standard boot-camp plot: a ne'er-do-well youth is, under the stern tutelage of a black drill sergeant, transformed into a real man, who earns his "whites" and the title of navy pilot. A second prefab plot formula is glued to the first as a poor but beautiful factory girl, decent and sexy, intelligent and caring, falls in love with the protagonist. Aiming to stay unencumbered, he behaves like a neurotic cad fearful of the "real thing"; however, he comes to his senses in the nick of time. The moment of epiphany arrives. In the film's concluding

---

* *Private Benjamin*, Wendy Chapkis and Mary Wings have concluded, is a "dangerous piece of propaganda . . . with a marginally feminist gloss."[17]

11

sequences, the freshly hatched pilot, now officer and gentleman, shimmering in his uniform, marches into the dull factory to sweep his sweetheart off her feet as her female co-workers applaud her victory: out of the factory, into the navy. When I saw this film at a packed theater in a shopping mall about three miles from the University of Massachusetts, Amherst, the mostly young audience (fifteen- to twenty-year-olds), pretty equally divided between males and females, burst into spontaneous cheers and applause as a grainy, soft-focus freeze-frame sealed the coming together of Warrior and Beautiful Soul. What were the young men cheering? Why were the young women weeping? What would it take to reactivate old symbols and reanimate old ideals, calling men to uniform and women to symbiotic service—this time in a cause we can win?

Although wars are primarily bequeathed to us as texts, war movies have a far greater impact on popular culture. John Wayne battling the Japanese on Iwo Jima; Audie Murphy going to hell and back; Gregory Peck stalwart through the heroic frustrations of Pork Chop Hill; working-class American men stricken and terrorized in Vietnam: these images reach millions. Although none of the immediate war or post-Vietnam–era films was explicitly "pro-war"—save John Wayne's moribund *Green Berets* (1968)—now we have seen released a cluster of blood-and-guts epics in which an unreconstructed loner, a Vietnam vet, goes back to settle some scores and rescue Americans still held prisoner. Gorily defiant, these screen heroes are winning *their* own war in Vietnam—and, in so doing, suggest the reconstruction of the warrior.

Public-opinion polls of relatively recent vintage simultaneously reassure and unsettle anyone concerned with war and the prospect of war. On the one hand, according to a *New York Times* poll, January 1985, only 19 percent of 1,533 respondents believed that we did the "right thing in getting into the fighting in Vietnam." But 68 percent of these same respondents expressed "a great deal" or "quite a lot" of confidence in the armed forces, and nearly half (47 percent) trust the government in Washington. The most trusting age group of all is the eighteen-to-twenty-two-year-olds.[18] As well, when asked if U.S. troops "should be used in a list of crises, support was higher in 1985 [as compared with 1974] in every case," including hypothetical invasions

Beautiful Souls/Just Warriors

of western Europe and El Salvador. The would-be warriors, aged eighteen to twenty-nine years, divide 55 percent to 37 percent, respectively, in favor of an invasion of western Europe should that be called for.

These several films and the poll constitute but a small portion of the wide-ranging evidence I shall tap on the durability of received cultural images of men, women, and war. Personal biography, political theory, historic case studies, feminist argumentation, and social history—these and more are grist for the mill.

13

1

Not-a-Soldier's Story:
An Exemplary Tale

Of course there are a good many times when there is no war just as there are a good many times when there is a war. To be sure when there is a war the years are longer that is to say the days are longer the months are longer the years are much longer but the weeks are shorter that is what makes a war. And when there is no war, well just now I cannot remember just how it is when there is no war.

—GERTRUDE STEIN
Wars I Have Seen*

THIS CHAPTER is intensely personal, my subjectivity and identity being the springboard for my discourse. I aim here to delineate, first, my encounter as a child and citizen-to-be with the larger, adult world of war and collective violence as it filtered down to me through movies and my family's experience; and then the witness I have borne myself, since my teens, as student, mother, and political theorist.

---

* History compounded Stein's irony, for *Wars I Have Seen* is labeled a "Wartime Book" and "Manufactured Under Emergency Conditions."

A Child of the 1950s: Images of War and Martyrdom

Approaching stealthily, leapfrogging from behind a bush to the camouflage of a large, gnarled tree, trying not to stumble over her weapon or to fall and skin her knees, or, worse, tear her dress, the determined, athletic eight-year-old prepares to surround her enemies—a curlyheaded, befreckled, slim six-year-old and a plump, red-haired, three-year-old toddler. The toddler has little understanding of the seriousness of the game. She wanders away from her ally, scampering in pointless circles on the lawn, prompting the six-year-old to complain to anyone within hearing distance, "Why does Bonnie have to be on *my* side anyhow? It isn't fair."* At that precise moment, the eight-year-old springs from behind her tree and screams, "On guard!"—an approximation of words heard in the Aggie Theater in Fort Collins, Colorado, earlier that day.

Her croquet mallet held aloft, the older aggressor challenges the six-year-old to a duel. "Come on, Patty, you're supposed to go on guard, too, and stand like this. Then we start to fight. Pretend you're hurt when I stab you." Patty, her croquet-mallet saber clasped in her right hand, shrieks and runs away. "Pat, that *isn't* the way it's supposed to be. You saw the movie. We both stand here and fight until one of us dies or gets hurt. Pat, come on. It isn't a game if you run away." Patty, screaming and racing wildly, shouts, "It's a stupid game," and, "You always win. Why do you always have to be the big boss anyhow?"

Bonnie, the red-haired toddler, recognizing the breakdown of the game, realizes she can do whatever she wants and flings herself to the lawn with gusto, then picks herself up and falls down again. Frustrated, the would-be duelist throws down her mallet in disgust, hoping it will break but it does not, and stalks off complaining that "nobody around here ever takes anything seriously." Storming into the house, the croquet-mallet swordswoman slams the back door, tears through the kitchen, and heads for the bedroom she shares with the three-year-

---

* When Bonnie got older, and could more reliably play an assigned part, she and I were often allies. In other words, I appropriated her and gave her orders. Power politics starts early.

old. Just as she contemplates whether slamming a second door is a good idea, a stern voice admonishes from the vicinity of the kitchen: "Jean, come back here. You go outdoors and walk in again decently. It's a wonder you didn't break the glass. Go on, I mean it." Humiliated, the girl retraces her steps, going out the back door and re-entering with elaborate mock courtesy, taking pains to inch the door shut so as to make no noise at all. Entering her room, she closes the door, lies down on the bed, and dreams dreams of derring-do. Swashbuckling . . . danger . . . handsome heroes and beautiful women. She gets to be a hero in those dreams, a woman in disguise, the greatest swordsman of them all.

On Saturdays, first came the chores and housecleaning. Everything had to be done uncomplainingly; there could be no hitch, nothing to jeopardize the promised trip to Fort Collins, twelve miles away, to see the matinée at the Aggie, sometimes the Lyric, theater. We kids had to earn a movie by good behavior. Mom didn't approve of movies in general, except for Walt Disney, Esther Williams, and anything to do with the life of Christ or Christianity (like *Quo Vadis*). She thought movies gave kids ideas and sometimes encouraged immorality by showing shocking things or not condemning strongly enough people who do shocking things.

My biggest movie adventure occurred on a Saturday sometime in 1948. I was eight years old and off to the movies alone, perhaps because Mom thought the film, *Joan of Arc*, might be too scary for my younger sisters. Swept away by Ingrid Bergman's Joan (see photo section, #1), her courage, her military prowess, her determination to remain true to her voices, her martyrdom, I left the theater in a determined daze and walked a block and a half to an establishment that was part barber shop, part beauty parlor. I told the receptionist I had come for a haircut with my mother's permission. Holding out a dollar bill and some coins, I said, "See, here is the money she gave me. I'm supposed to get my hair cut and I need to do it soon as I have to meet her at four o'clock." One of the haircutters, a young woman, being free, I marched steadily to the chair. "Does your mother just want your hair trimmed a little bit?" she asked. "Oh no," I replied. "My hair is too long. It gets in the way when I do my chores. I'm supposed to get it cut very short." "Are you sure?" she persisted, combing through my brown hair, which went halfway down my back. "Yes, it's supposed to be cut like this," as I simulated a brutal chop of all my tresses just below my ears. "And make straight bangs, too."

Reluctantly, the cutter began. I watched my long, shiny locks pile up on the floor and thought of seeing lambs sheared on my Grandpa's farm. I was scared. But it was too late now, and if Ingrid Bergman could do it to play Joan, I could do it, too. Never betraying my growing apprehension at how Mother would take this, I sat straight-lipped through the shearing, paid $1.25, or maybe it was less, and ran to meet her as I was already five minutes late. Her reaction was so bad I can't remember it. We drove back to Timnath, much of the time in stony silence. Twelve miles has never seemed longer. Dad would be hurt, Mom made that clear. And he was, bemused and sad at the atrocity his "Jeannie with the light brown hair" had committed on herself. While I cannot conjure up any details of the family furor, I can re-capture—as in an exemplary tale with me as dauntless heroine—my new sense of freedom, even though I had been quite vain about my hair. I did not care what the boys or the girls or the teachers said. One day I would be a leader of men, too. Maybe a warrior. Maybe a martyr—though there didn't seem to be much call for martyrs any more. I could worry about that later. In the meantime, I dreamed of action, of Joan, of myself in male battle attire, fighting for morally worthy ends. Oh yes.

Born in 1941, I knew the Second World War only at second hand, as did my immediate family. One of my father's brothers was called up, but my father was not because, as the sole teacher in a country school of grades one through eight, he had an occupational deferment. As well, he was married, with one, then two infant daughters. Most of the men in Dad's National Guard troop were in the war, and some of them died, in Italy, "at the Anzio landing." He still feels vaguely guilty about this, as if he should have shared identical risks and burdens with his friends.

Although Dad didn't see combat, he was in a war of sorts during the war. There was quite a bit of hysteria about anti-Americanism, and German "sympathizers" or German-speaking Americans were suspect, the latter often being assumed to be the former. Somehow a rumor got around that Dad was a German sympathizer not to be trusted with the hearts and minds of American youth. Eventually he had to go before some sort of school-board hearing to prove his Americanism and his loyalty, and even had to get the minister to testify to his character.

This story made a big impact on me, and I never forgot it. My mother's sister Martha and her brother Bill, in high school during the war, were also abused. Aunt Martha remembers her history teacher asking her each day: "Well, how are the Krauts doing today?"* And vivid still to my parents were First World War memories of the local Klan burning crosses on a nearby hill to protest German-speaking immigrants. War, I decided, was the occasion for outbursts of a debased sentiment I later came to call "nationalistic" or "xenophobic," reserving "patriotism" for authentic civic pride.

As a child I heard a lot about the terrible ravages of the Second World War from a friend of my mother's, a displaced person who, with her husband, had twice been made refugees—first after the Bolshevik Revolution and then after the Second World War. When Anna talked about the war, she would rock back and forth, and her voice grew singsongy and mournful, and sometimes she would cry, daubing her eyes with her apron. Mother said we Americans were "lucky" and could never understand how much the millions of DPs like Anna and her husband had suffered.

The first war I remember was the Korean, in 1952, when I was ten years old, and it intersected with the watershed event of my childhood—my contracting polio the year before. When I came home from the hospital, I was bedridden, unable to stand or to walk, and Mother had to nurse me full-time. The prognosis was not good—as I knew from the whispers I overheard at night beyond the closed door of the room made over as a sick room for me. But giving up meant you were weak and had no spunk, and I spent a lot of time thinking about Joan of Arc who wouldn't give up. I also remembered Abraham Lincoln's

---

* Aunt Martha later was a driver for the officers at a nearby camp for German prisoners of war, and there met her future husband, who was one of the officers in charge of the camp.

determination, learning to read at night in a freezing cabin by the flickering light of the fire and walking all those miles to return two pennies. He wouldn't give up either.

My devotion to Honest Abe had begun before the Joan of Arc revolt. But these two dreams—one of the warrior-soldier, a woman; the other of the war-hating, compassionate leader, a man—persisted side by side. After polio, Joan faded away. Warriors must be strong, able to run and leap and tear about. As I could no longer run, and was moreover soon caught up in the changes of adolescence, the story of Joan became a memory of childish romance and revolt; but that of Lincoln resonated through high school, into college, never deserting me, rekindled as I moved into adulthood and continued to struggle with what it means to be an American.

My mother's brother Bill served in the Korean War (she, I recall, was scathing about Truman's "police action"); and, though mostly in Hawaii, he did fly on a few missions, not in a fighter plane but as a crew member in a transport plane. Or so I remember. The highlight of my stay at Children's Hospital in Denver was the arrival—all the way from Honolulu, airmail and packed in ice—of a lei made of 150 baby orchids, together with a letter of endearment and worry from Uncle Billy. I requested and received a photograph of him in uniform, seated, smiling raffishly, his arms crossed, his cap akimbo. The nurses gathered round and raved about what a handsome uncle I had and how much he must love me and oh, how incredible to get all those orchids all the way from Hawaii. The nurses in Ward F told nurses in other wards. My bed at the end of the L-shaped ward was the site of bustling activity for several days as nurses flocked to look at my lei and to admire my soldier uncle's photograph. It went to my head, and I felt unusually favored, wishing Uncle Billy could be there in person to create a mass swoon among the nurses and to solidify my status. "There is something about a man in uniform," one young and friendly nurse said. "I don't know why, or what it is, but a uniform makes men look so handsome."

Back home, my head no longer in the clouds over my orchids, now dried but carefully preserved in their shipping box, I read the news-

19

papers daily and stared long and hard at *Life* magazine photos of the war. *Life* was especially enticing because among their correspondents was a woman, Marguerite Higgins.* In a khaki outfit, sharing life at the front, her blonde hair cropped sensibly short and appearing as a curly trim just visible beneath her hat, Higgins became a heroine to me. My dreams of warrioring dashed, it still seemed possible I might recover sufficiently to be a war correspondent. Reporters do not have to kill, and since others do not depend on them, they are not responsible for the deaths of buddies. Plus they can do a lot of traveling around in jeeps. Intrigued by Higgins and war reporting, I began to read war stories in earnest—not novels but reportage—discovering during the Korean War the Second World War's greatest war correspondent— Ernie Pyle, "the GI's correspondent." I checked all of his war dispatches, collected in three or four volumes, out of the Bookmobile. Trying to explain to my mother why I wanted to read all about war, I told her that Ernie Pyle "hated war but felt he had to share the risks with the men." He had the added qualification of martyrdom, having been killed in "the Pacific theater," I told Mother, "in the waning days of the war." A real tragedy.

Pyle is the chronicler of Everyman at War. His images are homey and haunting, and some I remember to this day, particularly his famous dispatch describing how the body of Captain Henry T. Waskow was carried down an Italian hillside by two of his men, struggling with the corpse of their dead and revered leader in the moonlight.† While I can't recall many heroics from Pyle, I do remember lots of stuff about homesickness and everyday ordeals with laundry, cold, mud, and K-rations.

Rereading Pyle recently, I gained renewed appreciation of his appeal. Pyle never celebrates war. He wages a war against war, against abstract celebrations of war and vague stories of campaigns that leave out individual men with real names and home towns and wives and sweet-

---

* In 1951, for her dispatches to the New York *Herald Tribune*, Higgins became the first woman to win a Pulitzer Prize for overseas reporting.

† I had forgotten Captain Waskow's name but looked it up in order to be true to the spirit of Ernie Pyle in writing this. For him the dead always had real names and came from concrete places, and he was determined to get the details right. In this light, it is ironic and touching that Ernest Taylor Pyle, Section D, Grave 108, SC3, USNRF, Aug 3 1900–Ap 18 1945, lies buried between two "unknowns" in the National Cemetery of the Pacific, Honolulu.

hearts. There is something so—*American* is the word that comes to mind—about Pyle's narrative style and persona: little tales of homey things. The heroics are those of men in often terrible and brutal situations trying to end the slaughter as quickly as possible. Pyle's GI Joes are not avid warriors in a classical mode but boys forced suddenly to risk their lives, who hope to survive and get back to the important business of normal living as soon as possible. This democratization of war and war narrative appealed to me when I was twelve years old, and appeals to me now even though I worry about that appeal.

I couldn't get enough of Ernie Pyle on the Second World War. I had, as I've already indicated, given up my own hopes of soldiering. But I had not relinquished my fascination with war, with combat, with tests of courage and loyalty. War promised a field of action more vital and serious than any other. War enlisted men in a common cause. Pyle brings such idealizations down to earth. After a while, the ordinariness of everyday army life, dominated by boredom, sleeplessness, and fear, took on an aura all their own, a kind of populist dignity very different from that evoked by George Washington rearing stiffly, all brocaded and bewigged, atop a monumental white stallion.

In demystifying war, Pyle involves his readers in it. The title of one of his collections of dispatches is *Here Is Your War*. And, in a democracy, it is "your" war. Pyle doesn't want you to forget that, to forget that you are implicated, too. It is *our* army fighting *our* war; and if we don't want the war, we shouldn't send young men to fight it, young men who would rather be home. Pyle writes:

> Was war dramatic, or wasn't it? Certainly there were great tragedies, unbelievable heroism, even a constant undertone of comedy. But when I sat down to write, I saw instead: men at the front suffering and wishing they were somewhere else, men in routine jobs just behind the lines bellyaching because they couldn't get to the front, all of them desperately hungry for somebody to talk to beside themselves, no women to be heroes in front of, damned little wine to drink, precious little song, cold and fairly dirty, just toiling from day to day in a world full of insecurity, discomfort, homesickness, and a dulled sense of danger.

The war is romantic to American soldiers "only twice," according to

Pyle, "once they could see the Statue of Liberty and again on their first day back in the hometown with the folks."[1]

Pyle's war narratives are stories not of glorious triumph but of occasional heroism, narrow escapes, and terrible sacrifice. What we see is war's simplicity, life pared down to basics, "food, sleep, transportation, and what little warmth and safety a man could manage to wangle out of it by personal ingenuity." The "usual responsibilities and obligations were gone, [with] no appointments to keep" and nobody caring "how anybody looked." There was "no washing of hands before eating, or afterward either. It would have been a heaven for small boys with dirty ears."[2] In this world without women, or with few women, "woman" becomes the desperately longed for symbol of home, femininity, warmth, gentleness, even as she also signifies the rule-governed existence of home ("wash your ears," "be on time," "wipe your shoes," "put on a clean shirt," "call if you're going to be late") that war gives men the opportunity to escape.

Despite Pyle's demystification of warrioring in the classic mode, his democratization of the soldier and "our army" carried its own potent appeal. Whatever was happening on the home front, what really counted was going on "at the front." That much was clear, and that was why to me Marguerite Higgins, breaking from the restraints of ladylikeness, signified daring. While army nurses were sometimes in the danger zone, too, they never interested me as a twelve-year-old because they were just doing women's work in an unusual situation— that is, being sources of comfort and succor behind the lines. More interesting by far was a Higgins—not only "at the front" but giving us what we had of that front. Pictures and stories *were* the war to me and to others.

To tell the tale gives power to the teller; he or she is implicated in the narrative and honored as a risk taker, for such one must be to tell this story. The woman on the home front just waits, I concluded, waits for letters, waits for the war to be over, waits for "her man" to come home, waits for a telegram she hopes never comes. The man comes home, and he has stories to tell, or so it seemed given the popularity of war movies and stories, but she has no story to tell because "nothing happened," no life and death conflict has challenged her. Maybe she puts together packages to send to soldiers, or becomes a frugal shopper, or struggles to tend to her family—like my mother and her sister Mary who as housewives didn't get, so far as I could tell, to do anything exciting. What were their lives compared with the moonlight on an Italian hillside and the drama of life and death? With coming under fire and learning whether one will panic or be brave? Or with making the Supreme Sacrifice? What, indeed.

Nobody in Colorado in the 1950s doubted that bearing arms was somehow a basic right. Whether America's romance with firearms has any connection to wars or militarism is hard to say. Societies that do not put handguns and hunting rifles into the hands of citizens by the tens of thousands also go to war, act jingoistically, and commit acts of collective violence. To the extent that gun ownership and use signifies male courage and helps constitute male identity, perhaps it does flow into those streams of thought and habits of being that more readily accept violence perpetrated by humans on other creatures, or on one another, as part of the human condition or at least of the masculine condition. What is being taught in the hunt is nothing like self-defense or just-war fighting, but stalking and killing an unarmed adversary who rarely fights back. Gandhi would see hunting as part of a larger derangement, the *dis*order of violence, and link it to wars.* Articulate defenders of hunting as sport insist that hunting is no preparation at all for war. The hunting narrative features an armed man against a "crafty" animal, preferably one "dangerous when it gets cornered." The aim is wholly individualistic, unlike the communal purposes of war. The modern hunter is neither protecting himself and his family nor putting food on the table. He hunts to test his skill and endurance and to bring home a trophy. Hunting, if it prepares a man for anything, prepares him for achievement in isolation.

Not as enduring and collectively inscribed as the war story, the tale

---

* As does Norman Mailer in *Why Are We in Vietnam?*—an unlikely pairing with Gandhi. Says a character at one point: "Maybe a professional hunter takes pride in dropping an animal by picking him off in a vital spot—but I like the feeling that if I miss a vital area I still can count on the big impact knocking them down, killing them by the total impact, shock! It's like aerial bombardment in the last Big War." Mailer's terrifying romp through the gun-laden, sex-obsessed, hunt-crazed mind of his protagonist, D. J. Jethroe, is a clinical dissection of one variant on the (male) American mind. The novel ends with D.J. and a friend going "off to see the wizard in Vietnam. . . . Vietnam, hot damn."[3]

of the hunt remains what it was in my childhood in Colorado, a peculiar marker of male identity. Although Dad succumbed to the strong social pressure in his community to hunt, I'm sure he never killed a deer in part because he knew his "womenfolk" would loathe rather than celebrate the achievement. Our cries of relief when he returned from the hunt empty-handed, acting rather disconsolate at yet another backwoods failure, fit into no accepted mold of females cheering male quarterbacks, or honoring great hunters, or valorizing the warrior. So no words were spoken; no antlers decorated our walls; and occasional jokes went round the table at family gatherings about my father's lack of hunting prowess. He took it in stride. How could he tell the other men that his failure enhanced his stature in our eyes? That would have made no sense at all. Mom was wrong: even Walt Disney movies give kids ideas. When Bambi's mother died, I knew hunting caused suffering. All the critiques of Disney's anthropomorphism miss the mark: on the deepest level, despite the cutesy images and the sentimentality, the story is true.

There is much mystery in the story of men and guns and forms of regulated or unregulated violence, individual or communal—as I understand in light of my own ambiguities and enchantments: finding hunting revolting and war intriguing, killing repulsive and guns fascinating. During my Joan of Arc period, I had begged for my own gun—a .22 rifle for target practice—having got the taste for shooting at a mountain picnic when I beat one of the boys. I didn't want to kill anything, save symbolically. But the idea of being a dead-eye shot and the image of going outside, rifle in hand: that intrigued. Keen though I was for a gun, my parents refused to let me have one. Years later, my youngest brother turned out to be a hunter, one who is skillful and responsible. I argue with him, but he never fights back. Hunting is important to him in ways I cannot understand. He is very much a young man who relishes solitude and is studying to be a doctor, a healer of bodies. He registered for the draft as a conscientious objector. There is "no sport" in war, he remarked laconically, and that was all he required by way of explanation.

Over the years I tried to make all this simpler, and sometimes nearly convinced myself that there was an unbroken chain stretching from boys' games to total war. I even felt comfortable, for a month or two, with the locution "men's wars" and the notion that violence emerges from the ugly recesses of the male psyche. But I knew that to be an evasion and hopeless as an explanation for the complexities of collective violence—as I had understood as a child back in 1952. And I am trying now to remember what I then knew.

Adult constructions of childhood memories are acts of the imagination—not a display of photographs posed with precision, but the recovery of fragile intimations. They do, however, figure in an evolving human identity. Sometimes our pasts yield smoothly into a clear present. For most of us, however, the course is stormier, and our own coming into being as adults a series of public and private, personal and political turbulences.

"Not-a-Soldier's Story" now moves, enmeshed in the filiations of childhood narrative, into representations that include journal excerpts and sometimes pained, more often ironic, commentaries on the complexities of identity and knowledge, of being a mother and becoming a political theorist.

The Growing Up of a Political Theorist

Monday, 31 August. The morning postman reported a great scare at Chelmsford on Sunday. An Ichabod telegram had been received there (founded on the reports with which *The Times* Sunday issue had been hoaxed, as I judge) telling that the British army had perished and that France was beaten. The "wire" was so full of despair that Chelmsford people could not take their tea. Happily, a little later, an official contradiction restored spirits and brought back appetite to normal conditions.
—Diary of the REVEREND ANDREW CLARK

The 31 August of this diary fragment was in 1914; and, from that year to 1919, the Reverend Andrew Clark logged detailed daily entries, saved memos, noted rumors, reported minute and major changes in the everyday life of his English village.[4] Clark—of Scottish ancestry, upright without being stiff-necked—approached his war diary as an assiduous believer might approach his or her sins: matters to be noted without embellishment, a faithful record serving as evidence of one's need for incessant discipline of the self. Clark was determined not to allow personal matters to fog his diary,* which he saw as a historical public record—in sharp contrast to my own strategy as a chronicler of my times, when in 1955, at the age of fourteen, I entered the era of daily journal keeping.

Thus, this second part of not-a-soldier's story is not, as in the previous section, a concatenation of dreamily remembered events but is specifically located, "pinned down" to a pre-existent text. Most of this material, written at fever pitch as I detailed public and private hopes and fears, did not survive a rite of passage in 1965, which took the form of a funeral pyre for ancient texts. But before performing this ritual, I read through my journals of 1955 to 1959 and culled select passages: the private historian at work constructing a narrative of selective remembrance. One principle of selection was whether an entry was more oriented toward public rather than exclusively private events. Teenage traumas of romance, sexual awakening, and struggles against parental restrictions got excised as too solipsistic, rather like an ache in one's limb, hence uninteresting to anyone but myself. (Also, I was embarrassed at my tendency toward romantic excess.)

Because of my concern with world events, my diary is riddled with wars and rumors of wars and evocations of different sorts of public figures. Unlike the Reverend Mr. Clark, I inserted my views on all these matters. His war diary seems to me remarkable by its omissions: a clergyman who never concerns himself with whether the Christian God might not be displeased at being called into service as Lord of war; who apparently had no qualms about the Church of England serving as an arm of state; who confronted no existential dilemmas about war or peace, life being a matter of responsibly filling out one's appointed vocation with honor and without fuss. I, on the other hand, a young American in a "rapidly changing world" (a locution drummed into our heads by our teachers who insisted that education was meant to prepare us to live in such a world), troubled myself at great length not only about what was happening but about what might happen and the "why" of both.

* He was so true to the narrative form he had set for himself that his beloved wife's death warranted no entry, emerging in the diary at second hand as Clark recorded the written condolences of others.

Not-a-Soldier's Story: An Exemplary Tale

There is little about the Bomb—it was difficult to believe in the nuclear threat if you lived in a small village in Colorado, trusting in both the President and God—save the single apocalyptic utterance of 14 June 1955: "There is just one word to describe war—hell. If we get in another war it will be the last one—there won't be anyone left to fight another." What inspired this doomsday sentiment I do not know. Although my Mom stored up bottled water and canned goods in the basement for a brief period during the height of the early 1950s bomb-shelter mania, she wasn't systematic and we weren't scared. Atomic bombs and Timnath, Colorado, just did not mix. The Russians wouldn't want us; and if they did drop a bomb on Denver, we would survive if we stayed covered until the fallout danger passed. Unlike other adults my age who tell me of their childhood bomb terrors, I never doubted that I would grow up and have a future. I never believed we would be attacked.

In June 1956, I celebrated the life and death of Nathan Hale, "truly one of the heroes of American independence." The famous last words of this young patriot, just before he was hanged as a spy by the British ("I regret that I have but one life to give for my country"), incited me to fever pitch for nearly a year. I had long cherished the theme of brave martyrdom for a cause and think I must have fallen in love with Hale, with the idea of Hale, with civic sacrifice and the poignancy of life cut down in its prime: he was, at the time of his death, only twenty-one and a young man of promise.

Throughout January 1956, when I might have been listening to Elvis with other teenage girls, I occupied myself by copying stanzas of nineteenth-century patriotic doggerel in Hale's honor as well as portions from an overwrought tribute to Hale by President Timothy Dwight of Yale College. The doggerel included an elegy by one Francis M. Finch:

> To drum beat and heartbeat
> A soldier marches by
> There is color in his cheek
> There is courage in his eye
> Yet to drum beat and heart beat
> In a moment he must die.

Another bad but wonderful verse, by one Virginia Frazer Boyle, ends with a rhapsodic evocation of a feminine American republic:

Oh, motherland, these are thy jewels, that blazon the shield of thy breast
Oh motherlove these are the truest—the hearts that have loved thee the
  best.[5]

Boyle was referring, of course, to the youthful patriot-soldier of whom Hale was an exemplar. Youthful idealism: courage in face of death; sacrifice for mother country; the bittersweet pathos of life tragically cut short. I read everything I could about Hale and announced to any who would listen that he had been given short shrift in high school history textbooks. Though I did not make the observation in 1956, now—rethinking Hale and the lore surrounding him—I find it striking that many of the remembrances were written by women. Those female versifiers set out, doggedly and didactically in the service of patriotism, to imagine a wartime event: overblown and stilted, their poems sink under the weight of patented metaphors and standard locutions. Though noncombatants, they mobilized language to sustain war stories.

Even as I dreamed a patriot's dream of Nathan Hale, I discovered a very strange book, an autobiographical account that embarrassed me in behalf of its author by the frankness of his disclosures concerning his own weaknesses and obsessions. All the stuff I was either afraid to write about or later excised was central to Mahatma Gandhi's *Story of My Experiments with Truth*.[6] His persona seemed alien in many ways. I did not understand the point of his "private" experiments, especially those involving *brahmacharya*, or chastity. But Gandhi as martyr and fighting man of peace, a warrior who did not kill but whose courage was never in doubt, exerted a powerful fascination. In August 1956, as a delegate from Larimer County to the Colorado State 4-H Club Fair in Pueblo, Colorado, I struck up a conversation with three IFYE (International Farm Youth Exchange) delegates from India, all men, all older than I (in their twenties). We talked about the Mahatma. They flattered me by declaring, "You know more about India than any other American girl we have met"—a statement I repeated over and over, to parents, classmates, teachers when I returned from the fair, aflame with my encounter with real "foreigners" with whom I had something—Gandhi—in common. My correspondence with one of the Indian delegates lasted about six months; my ambivalent relationship with Gandhi continues.

The twentieth of October 1956 found me worrying about the repercussions of Israel's attack on Egypt, calling the American position in all of this "precarious," though I cannot remember all the reasons it seemed so to me at the time. The portentousness of my entry for 4 November 1956 is signified by a note specifying the precise time I put pen to paper:

> 4:35 P.M. Russia has used and is using her military strength to crush and slaughter Hungary. The news we have received so far is spotty and confused. There has been no news from western reporters in Budapest since 5 A.M. this morning. It is not known what has become of them. The UN Security Council met in emergency session but a resolution to order Russian troops out of Hungary was vetoed by Russia.

Reports on Hungary follow for days, culminating with a sober musing for Sunday, 11 November:

> I wish we could have helped the Hungarians in their quest for freedom. . . . But for us to have intervened would have meant World War III and the horrors of thermo-nuclear war.

My last entry for 1956 reports on a "beautiful, wonderful, humorous, sad, happy, dramatic" film "about the problems that arise for a Quaker family during the Civil War"—*Friendly Persuasion*. I saw the film twice, maybe three times, each time wanting and not wanting the son in the family (played by Anthony Perkins) to take up arms for the Union and in defiance of his Quaker faith. Should he or should he not fight? I went back and forth on this one, but cheered unhesitatingly when Father (Gary Cooper) grabs his rifle and goes off in search of his son on the battle line. Other movie entries include ecstatic reviews of *The Young Lions*, with Marlon Brando as a German soldier; *The Diary of Anne Frank* ("a tremendous emotional experience"); and *Pork Chop Hill* ("one of the most realistic war movies ever filmed. The question was repeatedly asked: Just what are we fighting for?").

If I drew my sense of international politics from the daily news and *Time* magazine, my images of civilians and warriors in time of war came directly from the movies when I was a teenager. Years later, soldiers in Vietnam had a bitter name for one of their number who was too gung-ho: he was "pulling a John Wayne." Their sarcasm is further confirmation of the depth of the cinematic construction of American popular consciousness. Tragedy? *Anne Frank*. Courage in the face of apparent futility? *Pork Chop Hill*. Brutality and bonding between soldiers? *From Here to Eternity*. Nazi with a human face? *The Young Lions*.

Saturday, 19 January 1957, I tell a story about my brother Bill, then five-year-old Billy.

> Billy is such a character. Tonight he said to Pat, Bonnie, and me: "You're my girls and I'm your father."
> "Okay, Daddy."
> "Now, daughters. I have to go to war. Be good daughters while I'm gone."
> "Yes, father."

I offered no explanation for Billy's identification of fatherhood with marching off to war. Now, I suppose, one is called for, and mine will be cast in the form of what to avoid: war is there as imagery and myth, a pervasive figurative reality. To explain Billy's playacting with weighty psychological categories or knowing talk of conditioning and role modeling and socialization does damage to the exploratory nature of children's play in general and of my five-year-old brother's in particular.

A bright, observant kid, Billy clearly knew enough to know that war takes a man out of the house, perhaps for more than a few hours. He had no doubt picked up fragments of talk in preceding months about Israel, Egypt, Hungary, America, Russia, and how there "might be a war." Nonetheless, he was not on a trajectory toward a rigidly masculine/militarized construction of self. Billy, as a young man, opposed the Vietnam War and determined on a course of draft resistance, not draft dodging, should his number come up. (We were in the time of the lottery.*) He and Vietnam missed one another by a few months. My mother is still haunted by the sound of my brother's pacing at night, sometimes all night, struggling with his conscience, his citizenship, and his identity. He would not flee to Canada. He would go to jail if need be. Mom, angry about the war and frantic with worry, said the "whole family" would go with Bill to Canada. Not being a young, draft-age American male, she was not much concerned with the question of an honorable course of action. The immediate matter at hand, for her, was the safety of her son—saving him from the long hand of his, and her, government.

By 1960 my childhood was over. I was a college student and the young mother of an infant (my first daughter), and my examination of war and fears of war and male/female identity more and more fell to one side or the other of a line that severed official, public *discourse* from unsystematic, private *understanding*. The public student of history and politics, inhabiting the sphere of official public/academic discourse, being taught the ways of the political world as the "realists" (Machiavelli, Hobbes, Bismarck, Clausewitz) understood it, and the private dreamer, mother, novel reader, and Beatles buff parted company. In my journal I could sometimes heal the rift or draw the public/private selves into a relationship. Thus I recall a "horrible dream, a nightmare, of Nazi Germany. I dreamed that I tried to go into hiding (I was a Jew) and couldn't find anywhere to hide. There was more but I won't go into detail." There was no place for such political dreaming in the scientific study of politics as who gets what, when, where, and how; as *realpolitik*, a thick-skinned, skeptical, relentlessly unsentimental view of the world of men and nations.

Of *men* and nations: this remarkable locution made no impression on me at the time. Instead, I rehearsed the dangers of appeasement (the Munich analogy); the need for preparedness and defense; the difference between status quo and revisionist states; the disastrous effects of idealism in foreign relations as compared with the pursuit of national interest and *raison d'état*; why pacifism can't work; why international

---

* The draft lottery was instituted in 1969. Writes James Reston, Jr., "Service to one's country became a matter of bad luck." Reston also argues, and I agree, that defusing anti-war protest was one aim of the lottery system.[7]

relations and domestic policy cannot be run in the same way; what makes a nation-state and why nation-states are inevitable; the need to checkmate expansionist states before it is "too late"; why deterrence is necessary and the only chance for peace we've got; and so on. I could talk and argue like a seasoned *realpolitiker*, impatient with expressions of fuzzy idealists who were ignorant of the ways of the world. But I knew those ways through texts, through the filtering and appropriation of world events in a way consonant with the reigning paradigm of the academic discipline of international relations, or IR (see chapter 2, pages 86–91).

Text and the world collapsed into one another. I slotted most world events easily into a *realpolitik* frame, and dampened my skepticism with the meaning of "national interest" (What is it? Whose is it? How do we know it?). I placed my hope in vigorous, imaginative diplomacy. Kennedy was about to enter the White House. A new day was surely dawning. As the Cuban missile crisis came and went, and a nuclear-test–ban treaty between the two superpowers was signed, two new exemplars of moral courage and a politics of hope permeated my world, putting pressure on my newly hatched realism.

Martin Luther King, adopting Gandhian *satyagraha*, or "truth-force," to the struggle for black civil rights, preached militant noncooperation, peaceful struggle—fighting, yes, but nonviolently.* A politics stripped of moral consideration made no sense to King, no sense at all; and I knew it didn't to me either, not really, even as I mastered realist lingo and ways of thinking. Albert Camus's insistent focus on articulating limits to what human beings should not do even if they can—his arguments against terror and for solidarity, against nihilism but for revolt—helped me sustain a stubborn "other" to the *realpolitik* "self," as I hoped inchoately that I might one day put together mothering and political thinking rather than have to put aside the one in order to engage in the other.

Where was my voice? Was it a female voice, a mother's voice? Or that of a tough, no-nonsense expert, squeezing all possible sentimental nicety out of political thinking in order to bring our politics and our discourse about politics into a more approximate mimetic relationship?

---

* Paul Virilio might call this "holy *non* war" (see page 257).

Machiavelli didn't so much describe the world as it *is*, I thought, but as it must be to work in Machiavellian ways. If Machiavelli's is the only politics worthy of the name, and if we are to be political, it follows that we must make the world more Machiavellian, embracing a language of power, and instrumental violence and collective struggle, free from the stifling overlay of Christian or Gandhian or Camusian morality.

Torn between realist discourse and idealist principles, between strategic deterrence and civil disobedience, between the dominant image of the public man and the shaky vision of the private woman, her voice sounding strange and tortured as a public instrument, she—and here, aptly, I shift to the third person because I was very much at odds with myself—wrote papers about national interest and changed diapers and could see these activities only as the opposite ends of a pole cracked neatly in the middle.

In the future lay war and protest, private and public upheavals, feminism and a search for a female political voice. The entries that follow, unembellished, are my attempts in the thick of things in those heady years to think reflectively and ethically, often about political dilemmas; and my journal, and conversations with friends and family, were the only space for these attempts to clarify my point of view. I imagined an audience of sorts: sympathetic others with whom I hoped to have a dialogue; others who would understand first and judge later; others who did not think it crazy that I sometimes felt crazy, dazed and battered by public events even as my children grew, friends came and went, I earned my Ph.D., and life went on.

30 May 1962. A few notes on the movie *Judgment at Nuremberg*. The question raised by the movie concerns war guilt. Were only a few sadistic crackpots and SS men responsible for the horrors and atrocities of the extermination camps or were the German people as a whole not averse to such action and through their support of Hitler gave tacit approval? It is a difficult, perhaps an impossible question to resolve. One would like to think that only a few monsters were responsible but, unfortunately, it isn't that simple. According to Shirer, the German people had an authoritarian tradition and a tendency to be submissive to authority.* Nazism was an aberration of that tradition but it was supported by the German people as

* William L. Shirer's *The Rise and Fall of the Third Reich: A History of Nazi Germany* was a best seller and helped promote the thesis of collective guilt.[8]

a whole.* The businessman or the judge—seemingly good and honest people—through their support of Hitler and his methods and his aims paved a direct path to Auschwitz, Buchenwald, and Dachau. Of course, the responsibility of the free world in allowing Hitler to come to power cannot be abrogated. The policy of appeasement also played a part on the road to WWII oblivion.

War guilt settled, appeasement condemned, the "free world" spanked. At this point, when I was a junior in college, questions of the terror bombing of German cities and the use of the atomic bomb on Hiroshima and Nagasaki had not been addressed in any of my classes in a way that invited serious moral reflection and challenge to Allied and American war actions. I cannot recall any discussion at all, in fact, though I have vague recollections of hearing justifications for our actions, and those of our Allies, on the grounds of military necessity, war being hell. Truman acted to save a million American lives. Enough said; nothing debated—not yet.

Thursday, 2 May–Friday, 3 May 1962. Read complete text of Pope John XXIII's new encyclical, *Pacem in Terris*. It is one of the most inspiring statements of fundamental human rights and dignity I have ever encountered. He brings forth the concept—familiar to all who share in the Western tradition—of "natural law" and "right reason." He declares that every human being has the right to honor God "according to the dictates of an upright conscience." Further, each individual has the right to basic social services, to a living wage, and to an opportunity to participate in cultural affairs and the work of the body politic. Men† are by nature rational beings endowed with reason and free will. There is a moral order in the sphere of individual relations which should be carried over into relationships between states. The state exists to promote the common good and to respect fundamental rights. Racism and colonialism are condemned. Protection for minority rights is urged. Further, the Pope calls for disarmament, establishment of world community by strengthening the UN and an end to solving international problems by force or the threat of force. He reiterates a belief in gradual reform by working within established bodies politic rather than through violent upheavals. The encyclical is addressed to "all men of good will." I must think some more about his argument, especially

---

* This judgment must be much qualified: at least, this is the view emerging from the last decade of scholarship on the Nazi period.⁹ It was easier to be smug about this, and many other matters, in 1962.

† Before the days of pronoun debate and the notion of "inclusive" or "exclusive" usage.

his claim that domestic and foreign relations should be governed by a single moral law.*

Friday, 16 May 1963. Chanted by demonstrators: "I've never been in a concentration camp. Nor visited in hell. But I'm a Negro in Birmingham. And could describe both well." I wonder what it is like to be a member of a group consciously recognized as a "minority group," particularly a Negro. Those of us on the outside can never really know.

26 May 1963. Pope John XXIII, the Pope of Unity, Pope of Peace, died today.

6 July 1963. Saw *The Longest Day* this P.M. Very impressive film. But it seemed to lack grandeur or nobility of theme. I would not deny the realistic qualities of the film. It portrayed war as war undoubtedly is—dirty, grubby, frustrating. Death and dying in agony. A series of seemingly disconnected actions adding up to what is hopefully "strategy" or a Great Plan. The human blunders were clearly portrayed, especially those that were so disastrous to the Nazi cause. Excellent film, but I wasn't moved.

1 August 1963. In South Vietnam govt persecution of Buddhists exacerbates an already tense situation and makes abundantly clear the lack of support of the majority of the people in S. Vietnam for the Diem regime.

2 August. Diem is not only a dictator (perhaps a puppet dictator for his brother and sister-in-law) but a blundering one at that.... Undoubtedly, this lack of support for the regime, intensified by beatings, murder, and arrests of Buddhist monks and nuns and the imposition of martial law, hampers the fight against the guerrilla Viet Cong supplied from the North and Red China.

3 August. We should cut off Diem and his bunch now, I believe, and discontinue our support of a reactionary dictatorship.

3 November. Recent coup in South Vietnam ended the Nhu Diem regime and resulted in a military take-over. I wonder what role America played in this coup, if any. On the whole, liberal opinion, while wary of the military, welcomed at least the end of the oppressive ruling dynasty.

8 November 1963. Rdg Machiavelli. One especially good and entirely

---

* Reading *Pacem in Terris* was my first serious encounter with Catholic social thought—another of the many avocations I pursued as a *solitaire*, not from stubborn and willful isolationism but because I knew of no "community of interest" in the various settings in which I found myself.

true quote from the *Discourses*, Part I, p. 203. "Now in a well-ordered republic it should never be necessary to resort to extra-constitutional measures, for although they may for the time be beneficial, yet the precedent is pernicious, for if the practice is once established of disregarding the laws for good objects, they will in a little while be disregarded under that pretext for evil purposes."

From 22 November for six months, my journal is filled with little other than the assassination of President Kennedy and its public and private aftermath, including my decision to terminate graduate work in history with the master's degree so that I could devote myself more fully to politics and present concerns by going for a Ph.D. in political science. "Relevance" was the key term. Kennedy had called us to public life. By this time the mother of three daughters under four years of age, the closest I could come to answering the President's call was to switch academic fields! Although I "knew" better, the discourse of politics and political action merged in my mind. Somehow I would be more connected, more involved, less lost in a past time if I devoted myself to politics as constituted by academic studies rather than by history. My love for history, especially the Middle Ages, seemed, in the wake of the assassination, quirky and arcane, a romantic affectation, rather than a reflective affection. I remember muttering about the uselessness of knowing by heart the genealogy of the Merovingians and the Carolingians. I could leave history only by trivializing it.

Much of 1964 is fog to me now, and was at the time, as I attempted to hold together the pieces of a life unsettled in all its aspects. Although Vietnam was heating up, I burrowed inward, narrowing the circle, concentrating on private things—children, relationships, struggles with money and new living circumstances, trying to complete my master's in history before pursuing the Ph.D. in political science. I do recall President Johnson's speech after the Gulf of Tonkin incident, living-room discussions and arguments about Vietnam, and an early "teach-in" in which I, a part-time instructor in history at Colorado State University, participated.* I wasn't even sure what my position was

* The Gulf of Tonkin incident—involving two U.S. Navy destroyers that, according to President Johnson, were fired upon by North Vietnam without provocation—became the rallying point for Congressional support to deepen American involvement in Indochina.

going to be until the event got under way. I had prepared two sets of notes—pro- and anti–American intervention—and each set of arguments seemed equally compelling. When my turn to speak came, I grabbed my argument against American involvement and read it, although I referred to my pro-involvement remarks during the discussion when several of my allies began to indulge in what struck me as intemperate anti-American tirades, insufficiently wary of the North Vietnamese and their benign intentions. I could not understand then, and do not understand now, why so many in what became the anti-war movement coupled opposition to American intervention with celebrations of the glories of North Vietnamese democracy and virtue.

Why should those opposed to war on one level reaffirm it at another, deeper level by embracing the paradigmatic narrative of war discourse, including figurations of heroes and villains, good guys and bad guys, wicked imperialists and noble peasants? Shouldn't opposition to war deconstruct the narrative of victory and defeat? When I heard chants of "Ho, Ho, Ho Chi Minh/Viet Cong are going to win," I knew that the chanters had not repudiated war—they had simply relocated it—for they still dreamed of conquest. The war against the war was often warlike. Alienated from any politics that required that I hate, wrapped up in children and books and daily life, I wrote letters to the editor, walked in silent vigils on Sundays, shouted in horror at the television news, and threatened in despairing conversations to family and friends to relinquish American citizenship when *Life* magazine arrived, its cover a technicolor scene of massacred villagers, a baby about two years old prominent on top of the heap. I was ashamed of my country. It is terrible to feel such shame. Public dislocations tear at the heart of our identities. I could not believe that America had come to this. I took comfort in the Beatles, in John Lennon's playing the fool for peace and pricking all the pompous. I wondered whether I would survive motherhood and graduate studenthood. I puzzled over the celebrations the night Johnson said he wouldn't run again, for it seemed likely he would be succeeded by Richard Nixon. (I dared not hope that Bobby Kennedy might enter and win.) War entries in a journal I kept only sporadically in those years are few: the war by 1965—even before these dramatic events—had become a fact of daily life, the evening's television fare.

37

WOMEN AND WAR

2 January 1965. Saw superb film, *The Americanization of Emily*, this P.M. It pointed up the ludicrous nature of war. The dialogue was most intriguing. James Garner is the hero as non-hero and he expounds a kind of existentialism, I suppose. At any rate, he calls himself a coward: the religion of cowardice is what he practices. And if all the world were cowardly, he says, there would be no war. He wishes to know himself as he is—Truth he leaves to God. His value system goes from the woman he loves, to his home, to his country, to the world, to the universe—in that order. In the end he has to sacrifice certain values, i.e., his hatred of the glorification of war, for the dearest—life itself and the woman he loves. A touching film: comic pathos, an intermingling of human foibles with flashes of insights and love and reality—war, for example, is not so much attacked but hero worship of the bloody mess is.

25 December 1969. The Kennedys are dead. King is dead.* Optimism is dead. America is dead. (It died in Vietnam.) We have no illusions left. Few illusions can possibly remain to those of us who entered the decade not yet 20 years of age [and] have spent at least ⅓ of our lives in years that saw the creation of H-bombs, MIRVS, germ warfare, discovery of the extent of pollution, etc.†

3 May 1970. Heidi [my nine-year-old daughter] had a nightmare. She dreamt bombs were falling on our house. We were being bombed from airplanes the way our airplanes bomb Vietnam. She said she tried to think of "April showers and spring flowers, but the flowers turned into bombs." [In Ingmar Bergman's masterpiece *Shame* (1967), the dying Eva, hopelessly adrift with other refugees, including a husband from whom she has become estranged, wearily recounts a dream, in which there are beautiful roses— but they are all burning.] Damn this war. How frightening and terrible that in so-called peace-time (for Americans) our children dream of bombs.

28 December 1972. Harry Truman's funeral just started. I'm not sure about Truman. He may have been directly responsible for more human deaths than any other American President, though this sounds banal the moment one notes it. The deaths must be placed in context to make some

* Martin Luther King was shot and killed on 4 April 1968. Senator Robert Kennedy was shot two months later, on the night of 5 June, and died the next day.
† This entry is a reminder of a moment of collective depression, a time in our history when, because politics was going badly, "not only disappointments but also dislocations are likely to result": the words are Michael Sandel's. According to Sandel, in *Liberalism and the Limits of Justice*, "when politics goes well, we can know a good in common we cannot know alone." But we face the dislocations alone for chief among these is isolation, a decivilizing (in the civic sense) of the self.[10]

sort of historic judgment. I am offended by the sloughing off of Hiroshima. Even if one ignores the revisionists who argue Truman dropped the bomb to scare the shit out of the Soviets, there are problems. Let us assume Truman really *believed* the bomb would save some amazing number of American lives—1 million is the figure usually bandied about. Following Truman's justification it follows that one must adopt one of the following possible calculi: (1) It is better to kill 100,000 for sure than 100,000+ almost certainly, or (2) It is better to kill Asians in any number than to risk the deaths of Americans ("our boys") in any number. There is an important difference between (1) and (2). Calculus (1) is the sort of statement even Christian statesmen [if those two terms aren't mutually exclusive] must apply given the vagaries of a system of international anarchy. It differs in magnitude but not essence from the calculus a "normal" person would apply in a situation in which he sacrificed "the lesser" for "the greater." But calculus (2) is insupportable. It is lodged, not in respect for *human* life, in which case one wishes the deaths or injuries of as few as possible, but in ethnocentrism, if not racism, holding the lives of one type of human (white American) dearer than those of another type (brownskinned, non-American). Is this calculus "allowable" from any ethical (as opposed to straight *realpolitiker*) vantage point? I think not. . . . It is sad. Truman himself said it, I guess, that he never "looked back": he made his decision and, as they say, that was that. He didn't ruminate, didn't allow second thoughts, doubts, or remorse. "The Buck Stops Here," "If You Can't Stand the Heat Get out of the Kitchen" and all that drivel. 100,000+ incinerated Asians not worth a backward glance? Well, Truman believed "it" was necessary to save lives. But when the evidence began to suggest this might not have been so did he, in the still of some Missouri night, wonder whether he might have held off a week or two? As he walked the halls of his Library when it was closed for the day and he had it all to himself did he ever put *himself* in Hiroshima? Did he ever question whether one man should have the power to drop such a bomb on his own authority no matter what the situation? I suppose not. That's probably why they say the Presidency didn't age him.

9 May 1972. I was working. L. phoned. K. had terrified her with talk about what-is-going-to-happen-now in Vietnam given our noble leader's mining of the harbor of Hanoi and the revelation of the fact that there is a time bomb in Haiphong harbor set to go off at the moment Nixon's deadline of last night (3 daylight days, i.e., by Friday morning) is reached.* There are 19 ships in the harbor now. Presumably they will all be obliterated

---

* K. and L. were close friends concerned about the terrible things that seemed to be happening daily, and sometimes turned to me as an "expert"—the political thinker.

unless they quite literally "ship out." K. told L. a nuclear holocaust was coming—that Russia wouldn't stand for her ships being blown away and would retaliate by dropping a nuclear bomb on Saigon or something.

I pointed out confidentially and reasonably (never know when a seminar in International Relations will come in handy) that the international system is a low-reactive system; that neither Russia nor China nor the U.S. wants a major war; that neither Russia nor China, in the last analysis, give a shit about Viet Nam per se and if the crunch comes they will abandon Viet Nam rather than risk their own national security and well-being and, moreover, that there is no way they could get a nuclear payload to Saigon given our air and sea superiority in the region. So I did all this rational chit-chat replete with "facts" and "figures" and how Nixon, as disgusting as he is, is not a madman itching to put his finger on the nuclear button. He has a Great Concern for his Image in the Eyes of Posterity. This means there must be a posterity with eyes to see. For this reason if no other he would not want the rapprochement to China overshadowed, as one could reasonably predict it would be, by a nuclear holocaust.

So I told L. to listen to some Beatles records and stop worrying. But when I hung up I realized I felt a chill, a clammy sense of dis-ease. A nuclear war *is* possible. And now I find I can't pick up the thread of my pre-phone call thinking on the problem of female guilt, an overworked topic in any case. Postulating the end of life on earth rather overshadows all else.

As my children grew, so did my war worries. The public and private strands drew closer, eventually blending, becoming that which I think and write and teach about as well as that which I fret about as a mother and citizen. Seeking in the experience and way of being of mothering some alternative to militarized identity and discourse, I entered the terrain of feminist debate, discovering among many feminists an animus toward a maternal woman's voice as pronounced as that of any *realpolitiker* male from seminars in the past. Even as I experienced the sting of public repudiation of my arguments by other women, I gained an appreciation of my own mother's blunt wisdom, recalling her words about Vietnam and my brother Bill: "I didn't raise him to be cannon fodder. If the country were really threatened, the mothers and sons could fight together. But he isn't going there. I won't allow it." At the time I admired her spunk but faulted her reasoning: it isn't that simple, Mom. There are considerations x and y and z. Of course. But is there not basic justice in her remark—a recognition that all should be prepared to defend, should it be imperiled, a way of life they cherish; but that none should be called upon to die in a cause that inspires as much fear and loathing as it does valor and commitment?

I'm not sure about mothers and peace and "maternal thinking" in the public sphere. But I do know that I will never be the mother pictured in a Second World War–vintage newsreel, shrieking with pride and joy when her son's name is pulled from a jar, giving him the honor of being the first drafted. I can more easily fathom, and respect, the mentality and identity of the warrior than that of the members of a 1930s organization called Gold Star Mothers of Future Wars, women anticipating the deaths of sons yet to be born. If I were the son of a Gold Star Mother of a Future War, I might have a qualm or two about whether my yearning for longevity and my mother's quest for valorized warrior-motherhood might not one day come into conflict.

If the mothers refused to let their sons go, wouldn't that stop things? Of course, Mary Gordon is right: it is sentimental even to pose the question. But it is defeatist and immoral not to pose some such question. My son turned eighteen on 5 May 1985. He has been worried about war and the draft—explicitly talking about it—since he was twelve. He is angry at the perceived injustice of having federal student loan money withheld as a penalty should he fail to register. He is short-tempered with "girls at school who tell the boys what they ought to do, or what they would do if they had to register, when it's nothing they have to deal with so it's annoying for them to go on and on when they are not coerced and they don't face risks and penalties or have to worry about whether their motives are really decent or just selfish."

Eric Paul Elshtain received his registration acknowledgment letter in the mail a few weeks after his birthday and his troubled decision to register. He was advised to keep the letter as "proof of your registration with the Selective Service System." He was reminded of all the requirements of federal law—notifying Selective Service within ten days of errors, changes in address, name changes, and so on—and warned that "failure to comply might result in a $10,000 fine or five years in prison or both." An "important note" was appended: "Certain Federal and State legislation has been passed that requires registration-age men applying for benefits (such as, Federal student financial aid) to be registered with Selective Service." And, as a friendly civic tip, Eric was admonished not to forget to register to vote.

His ten-digit Selective Service number now part of a great whirring computer somewhere in the bowels of the federal government, Eric added under his name and address: "Registered as a Conscientious Objector." When he had left the room and could not see me, I allowed the tears to flow: five words, not much protection should the machinery be set in motion. A young man whose first political awareness was, at age seven, passing petitions to save whales; who became a vegetarian because he could not justify the manner in which animals are raised and killed for human palates; who has never been in a fistfight in his life; who worries about pushing people around; who spends hours with his older, mentally handicapped sister, listening to her halting speech, helping her with daily matters; who wants to write poems and short stories—it would be a crime to call upon one such to kill. It seems criminal to put lethal weapons in the hands of *any* eighteen-year-old—but the requirement to kill is more debilitating for some than for others. His mother says she will never allow it. The political theorist, Professor Elshtain, knows she cannot stop it with argument, but she knows no other way to try.

Torn between commitment to my own anti-militarist male child; fearful, too, for my daughters and my nieces and nephews; mindful of all the arguments concerning citizen duty and obligation and commitment to the common good, I end for now not-a-soldier's story as a narrative still in the making, as I search for a voice through which to traverse the terrain between particular loves and loyalties and public duties. The stakes are terrifyingly high—no longer fantasies of the use to which I might put my own body, as Joan the warrior or Marguerite the correspondent, but the perilous possibility that others might, in Mary Kay Blakely's words, "come calling for the bodies of the children"[11]—now both male and female.

This struggle is not war in our received triumphal sense, or even our modernist ironic and cynical sense; but it is a fight, a fight that pits the moral voice against the insistencies of statecraft: Antigone everlastingly against Creon; a disillusioned pacifist Jane Addams against a John Dewey reconciled to war; Dorothy Day against Cardinal Spellman; politicized women, most of them mothers, against political men, most of them fathers. But the matter is not so clear-cut as Antigone versus Creon. As George Eliot knew:

> Reformers, martyrs, revolutionists, are never fighting against evil only; they are also placing themselves in opposition to a good—to a valid principle which cannot be infringed upon without harm. . . . Wherever the strength of man's intellect, or moral sense, or affection, brings him into opposition with the rules which society has sanctioned, *there* is renewed conflict between Antigone and Creon.[12]

In Howard Hawks's 1941 classic *Sergeant York*—a film celebrating a Just Warrior of the First World War in anticipation of the Second World War, marking the transformation of Alvin York, played by Gary Cooper, from Biblical pacifist to dutiful citizen, loath to kill but recognizing that each must defend the rights of all (as the film puts it and York endorses it)—the womenfolk, simple backwoods Tennessee countrywomen, stand mute and perplexed as York is drawn from his rural home, to go off to be trained as a soldier and sent across an ocean to fight. When Alvin has said goodbye, his mother and sister, their eyes following him down the path away from their humble cabin, have the following conversation: "What are they fightin' fer, Ma?" asks the daughter. "I don't rightly know, child, I don't rightly know," says Ma. Not rightly knowing, she can offer neither resistance nor affirmation of her son's decision to fight. It is all a mystery to her. While realistic to backwoods rural life in 1917, "not rightly knowing" no longer rightly suffices. Simple "fers" and "agins" are the stuff of homespun yarns. If soldiers increasingly demand reasons and justifications for why they must fight, not-soldiers—sisters, mothers, wives, and lovers of soldiers or would-be soldiers—must make similar demands and resist the parts to which they have been traditionally assigned in soldiers' stories. To assay those parts, I turn to exploring the discourse of war, searching for the roots of an inheritance that features civic virtue as armed and ready to do battle.

PART I

ARMED CIVIC VIRTUE

2

The Discourse of War
and Politics:
From the Greeks to Today

I N THE BEGINNING, politics gave birth to war. Better: in the beginning, politics *was* war. The story of politics and war in the Western tradition does not unfold as a fall from grace, a tale of sordid descent from a bucolic age when people peacefully went about their business and let their neighbors peacefully go about theirs. Instead, it is a tale of arms and the men. Long before there were Just Warriors and Beautiful Souls—a construction that emerges with the triumph of Christianity—larger-than-life exemplars of brutality and revenge, of martial and maternal honor and civic peace, bestrode the pages of tragedies, formed the stuff of myth, and congealed into durable legacies, sublime and terrible. For the Greeks, war was a natural state and the basis of society. The Greek city-state was a community of warriors: "Political leaders were military leaders. Political life entailed the preservation of the city through war."[1] The funeral oration of Pericles enshrines the warrior who as the true Athenian has died to protect the city.[2]

Historians of the classical world note a "direct line of descent from the Homeric warrior assemblies to the Athenian naval democracy."[3] The Greek citizen army was the expression of the *polis*; indeed, the

47

creation of such armies served as a catalyst to create and to sustain the *polis* as a civic form. In Sparta, much admired centuries later by Jean-Jacques Rousseau for its robust unity, the army was familialized as the basic unit of, and foundation for, the city-state; and the male custom of homosexuality, there as elsewhere, served to cement bonds between those who would likely fight and die together.

Important differences separate the mythic ideal of the Homeric warrior from the citizen-soldier of institutionalized civic life. The tales told of these distinct forms of warrioring signify far-reaching transformations. The great tragedies make way for the history of Thucydides and the political theorizing of Plato and Aristotle. Shifts in genre mark transformations in the structure of Greek experience. With the triumph of the Greek citizen army and the city-state, war is less akin to the *force majeure* of the classical tragedies—an impersonal yet living "it," a swift and sudden flash flood sweeping human beings along in its frenzied wake—and more a matter of political decision making, rules, and strategy.

The story I aim to tell is not a chronological history of war and politics, or politics-as-war from the Greeks to us; rather, I am interested in the ways war stories are deeded to us as texts of a particular kind. Narratives of war and politics are inseparable from the activities of war and politics; each—writing about and doing war and politics—are practices existing in a complex, mutually constitutive relationship. I espouse no vulgar notion of mimesis here. Rather, stories of war and politics structure individual and collective experience in ways that set the horizon for human expectations in later epochs. Nancy Huston puts matters forcefully: war imitates war narrative imitating war.[4] To cast another spin on it, the world Niccolò Machiavelli described "realistically" in the sixteenth century was a world cut and measured to fit his assessment of his political culture and his remedy for its travail. "Realist" diplomacy and statescraft, taught to generations of students of politics as the "scientific" study of politics, takes its starting point and its central lines of demarcation from a Machiavellian narrative, a story of civic virtue, its risings and fallings, a creative interpolation that is in fact sur-real, a construction of and gloss upon perceived reality. The politics of the text distorts by expressing exaggerated fears and hopes—amplifications that go on to become embedded in practices.

In taking up the discourse of politics and war, I have been startled at the continued sway of received narratives and reminded of the genealogy of potent concepts: power, force, civic autonomy, *raison d'état*, civic virtue, patriotism, honor, glory, necessity, tragedy. There are several possible ways to traverse this territory. I begin from the standpoint of the present, immersed as I cannot help but be in the "to" that our discursive "froms" help constitute: from Machiavelli to MAD ("mutual assured destruction"), from Rousseau to Reagan. The prepositions define a trajectory, "to" signifying the present truth claims of a historic discourse.* That discourse, political theories of war and politics, is not *of* a piece but might be visualized as *pieces* of a gigantic puzzle that somehow fit together or have been gerrymandered (an edge smoothed over here, a sharp defining angle serrated, an unseemly protuberance hacked off) to fit. In this chapter, and the next, I shall put some of the pieces together in order take the measure of a tradition I tag "armed civic virtue."

The Greeks are the "from" that sets in motion the "to" of the present *if* you are approaching questions of war discourse as a political theorist. They invented political theory and politics, shaping tropes, offering metonyms, articulating various problems for thinking about and doing politics, that continue to spook us at this late date. While we cannot avoid being haunted, we can try to tame the ghosts and specters.

Taming Homer's Warrior: Plato and Aristotle

For Simone Weil, Homer's *Iliad* is a poem of force, a relentlessly grim account of the way of being of an aristocratic warrior class for whom war is natural, honorable, glorious, and tragic. Cities are sacked, slaves are taken, guts are spilled, eyeballs are wrenched, pleas for mercy are rejected, glory is attained, honor enshrined, defeat inflicted. It is a story for the stout of heart.[6] Nancy Huston suggests that the mourning of women, their tears and lamentations, is "one of the goals of war"

---

* Dennis Porter has helped me sharpen this discussion.[5]

in the *Iliad*, not one of its "unintended consequences," to use contemporary jargon.[7]* The key words in the warrior's tale are *glory* and *honor*, and honor lay in physical courage. Steven Salkever notes: "The Greek word most frequently used to express this quality is *andreia*, the virtue of the *andres* or real males, a word ordinarily translated as 'courage,' but perhaps more tellingly rendered by 'virility' or 'manliness.' "[9] He adds that manly valor is shown by the spirit, or *thumos*, with which men pursue honor and fame. Women inhabit a different field of honor, the household. The woman who acts violently (in "male-like" fashion)—for example, when Clytemnestra murders her husband, Agamemnon—the response (of the chorus in Aeschylus's *Oresteia*) is horror. Male revenge and killing is public, sanctioned, has a larger purpose. Female killing is disorderly conduct, private revenge spilling beyond the bounds of the household, or *oikos*, to threaten the bases of social order (I discuss female violence further in chapter 5).†

The tradition of the great epics and tragedies—including Sophocles' *Antigone*, Euripedes' *Trojan Women*, the *Oresteia*, and the *Iliad*—are backdrop to the emergence of political theory; as, too, are Aristophanes' *Lysistrata* and *Ecclesiazusae*, comedies in which women, by politicizing and extending their domain, repudiate the male world of war and its terrible dislocations. The women's solution to war/politics is familial society writ large, not so much an alternative vision of what politics might mean as the eradication of politics for an encompassing domesticity.[11]‡

Representations of female warriors or mythic female-controlled so-

---

* From a different angle, Hannah Arendt might agree. Although she doesn't zero in on the women of the Greek epics—they exist in a private sphere that is hidden and wordless in Arendt's account—the point of male action is to attain immortality, to demonstrate a "capacity for the immortal deed": hence, Achilles.[8]

† Arlene W. Saxonhouse has a splendid discussion of the *Oresteia* that puts Clytemnestra in a positive light.[10]

‡ Antigone refuses the truths of war, or seeks to circumscribe them, and, in so doing, sets for ensuing centuries the conflict of private conscience versus statecraft. Recall the story: the *dramatis personae* that matter for my purposes are Creon, King of Thebes, and his nieces, Antigone and her sister, Ismenê, daughters of Oedipus. In the higher interests of the state, Creon issues an order that violates the sacred familial duty to bury and honor the dead. Antigone, outraged, defies Creon. Ismenê argues that women cannot fight with men. But Antigone, determined, insists that there are matters so basic that they transcend *raison d'état*, one's own self-interest, even one's own life. George Steiner argues that in Antigone one finds, in his words, "a fragile intimation of humanistic ideals" with Antigone as a summoner of "futurities of conscience."[12]

cieties also titillated the ancient imagination, the most fearsome being the Amazons whose self-mutilation signified the repudiation of the "feminine" their warrioring demanded.[13] The goddess Athena is a virgin born from a man, emerging from Zeus's head armed and shouting a war cry.* But these potent figurations, for all their fascination, do not detract substantially from the central theme: that warrioring is a male affair. The maternal Hecuba mourns but acquiesces in the bloody business, advising Andromache, her daughter, "Give your obedience to the new master; let your ways entice his heart to make him love you. If you do it will be better for all who are close to you." Mourning the death of Andromache and Hector's son, her grandson, Astyanax, Hecuba decries his "wretched death," noting, "You might have fallen fighting for your city."[20] *The Trojan Women* presents "the model of the epic hero who can only find fulfillment through the acquisition of glory attained in the battlefield. . . . And it is this model which has

---

* This probably isn't quite fair to Athena. She is a warrioress *and* a force for peace as well. Although she emerges in full-blown battle gear from the head of Zeus, she often lays aside her helmet. In the *Eumenides*, she turns conflict into peace and reclaims her maternal roots, giving birth to Athens. The *Eumenides* concludes with Athena calling upon the Furies to undergo a *civic* transformation: from "daughters of the Night," representing the old blood feuds, to become cherished civic spirits, to sing a song to bind the land forever, a song, in Athena's words, of "only peace-blessings, rising up from the earth and the heaving sea, and down the vaulting sky let the wind-gods breathe a wash of sunlight streaming through the land, and the yield of soil and grazing cattle flood our city's life with power and never flag with time."[14]

To be sure, earlier she proclaims that "our wars" should "range on abroad," but she fights "the curse of civil war."[15] Nancy Hartsock's reading, in *Money, Sex and Power*, eliminates Athena's doubleness, taking up sides with the Furies and construing them as female forces that must be brushed aside in favor of the claims of father/son and "male supremacy in all things."[16] But this doesn't really work. First, because the Furies are not simply externally threatening female forces that must be crushed: they are *within* Orestes, crying for blood. The Furies embody both negative and positive extremes, within and without. They are, as Robert Fagles argues convincingly in the introduction to his translation of the *Oresteia*, both Orestes' punishment and his power. The Furies are not so much female emissaries being crushed but forces *within* each being that simultaneously push toward blood vendettas *and* social justice, a law of retaliation and a law of regulation of the city.[17] Second, Hartsock quotes Athena but alters her speech judging in favor of Orestes with an ellipsis that distorts Athena's meaning. Hartsock cites Athena to the effect that she will support "the father's claim/and male supremacy in all things," and, in a footnote, adds that Athena makes "one exception to this support of male supremacy—the giving of herself in marriage."[18] But if one takes Athena's words intact, something very different happens: "I honour the male, *in all things but marriage*."[19] In this way, Athena lends her support to a *civic* institution, moving away from the blood ties of the kin/tribal order. In marriage her loyalties will not be exclusively with the male. By insisting on the essential ambiguity of Orestes' actions, urged on by the/his Furies, Athena can help the Furies find virtue in their vital energy, turning it from vengeance to life.

caused the women so much suffering that they nevertheless continue to accept."[21] From their standpoint in the Homeric age, women do not challenge war's inevitability, though through the eyes, ears, and voices of a few representations of women, its glory is tarnished. It is this world, with its backdrop of brutal glory and heroics, that Socrates undermines, in favor not of pacifism but of an armed civic alternative.

The dramatic transition from the world of an Agamemnon or an Achilles to one in which Sophists preach and practice rhetoric, assemblies of citizens debate, and philosophers interrogate, has a resonant analogue in American cultural experience. The classic Hollywood western embodies an orientation to violence and institutionalized civic life that speaks to the incompatibility of a highly agonal warrior ethos with the collective lawlikeness and organized defensiveness of the town or the *polis*. Themes and identities given us by the Greeks are reflected in our dramatic tales of rival claims and conflicts. Thus, the prototypical hero-protagonist Shane (from the 1953 classic of the same name), an ex-gunslinger, helps create the possibility for law and order by rousting the ranchers and their hired gun. The ranchers are exemplars of violent self-help—anarchic, thuggish warriors.* Their penchant for cutting down the fences of the settlers, trampling their gardens, and running their domestic animals—before they move to more drastic measures if these attempts at intimidation fail to do the trick—expresses their motto: "Don't fence me in."

Shane, who has put up his gun, is recalled to his violent vocation when the peaceful settlers find themselves outclassed by a hired professional warrior brought in by the leader of the ranchers. (A settler is by definition one who is "settled," who stays in place, seeking a rule-governed, unharried existence without need for grand heroics.) When the settlers demonstrate their inability, even banded together, to face this escalation of violence, and after one of their number has been ruthlessly gunned down, Shane reluctantly rearms one last time—for the good of the collective, yes, but for individual glory, too. Having done the violent deed, he recognizes that his continued presence in

---

* Should anyone be unfamiliar with this great film, its date is 1953; its director, George Stevens; and its stars, Alan Ladd, Van Heflin, and Jean Arthur. The young boy's plaintive cry in the film's closing sequence—"Shane, come back, Shane!"—expresses a loss at once individual and collective.

the community would be a threat to the civic peace he has helped create. Shane *must* ride away at the film's conclusion. He is as out of place in the placid, settled kingdom of a tamed town on the American frontier as was Achilles' gory glory in Aristotle's list of civic virtues. *Polis* and *polite* are etymological cousins and are neither the setting nor the evaluative term one would pick for an Agamemnon or a Hector.

When Socrates steps onto the stage, a new discourse—cast in the form, first, of Plato's dialogues, then of Aristotle's sustained analyses— takes shape as the pre-eminent narrative of and for the city, the real and the ideal *polis*. The shift thus marked as the *polites* supercedes the warrior is not one from war to peace. What is at stake instead is collective understanding of war as a human undertaking and an object of reflection. Recalling the dismemberments, individual and collective, of the epics and the tragedies, Plato seeks order, and the Homeric warrior ethos is challenged as a vision of male honor. War may bind the body politic, but only as a regularized collective activity, undertaken from civic necessity, not from an individual search for glory or vengeance or lust to annihilate. There is no room in Plato's *Republic* for the war lover. But there is ample space and need for the Guardian- warrior, a zealous defender of the autonomy of the city, wholly devoted to its good and cut off from "selfish" pursuits.

War as a pervasive necessity remains, but the aim of fighting is to preserve the collectivity. Alternative visions of honor and virtue are part and parcel of Plato's philosophy, a complex textual articulation that simultaneously affirms and challenges features of the Athenian social order. The Homeric warrior ethos does not emerge unscathed as a sufficient vision of male (and female for Plato's Guardians) honor once Socrates has worked his way with his interlocutors in *The Republic*.* Important for subsequent images of armed civic virtue is Plato's insistence that those who aim to found a city must reject the "models" of the poets by refusing to sanction tales that incite unruly passions, sexual and aggressive. Homer takes his lumps.

There are, however, appropriate civic myths, ideals that hold the city together. Several customs of the Homeric warrior group are re-

* Plato's elevation of a few select women to the level of Guardian is best explained as part of his insistence that collective *esprit* is maintained when the "best" rule and share everything, including women, in common.[22]

tained—for example, the tradition of taking meals in common—even as the ideal of male virtue embodied in the warrior is deflected. Plato's Guardians must "go regularly to mess together like soldiers in a camp and live a life in common."[23]* In his thrusts and parries with his less nimble verbal sparring partners, Socrates adumbrates rules for civilized warfare: the terms under which enslavement of the vanquished is permitted; whether it is not "illiberal and greedy to plunder a corpse," or "the mark of a small, womanish mind" to treat the body of "the dead enemy" as spoils of war. The tension in Plato's position is that his city requires a trained, disciplined, unified class prepared to govern wisely in peace and to fight courageously in war. But because he sketches the traditional warrior as a rather thick fellow with fixed dispositions—you can count on him as trustworthy, but he is "hard to move and to teach" as if he "had become numb"—*his* warriors must be philosophers, too.[24]

Further tempering of the warrior ideal occurs in the work of Aristotle.† His *Politics* is critical of "the idea that virility or courage is the foremost human virtue ... and the Periclean opinion that all quiet people and cities are useless." He criticizes those Greek city-states whose politics aims at domination over others, a goal that necessarily elevates war over all other forms of activity. Sparta is taken to task in *Politics*:

> The whole system of legislation is directed to fostering only one part or element of goodness—goodness in war—because that sort of goodness is useful for gaining power. The inevitable result has followed. The Spartans remained secure as long as they were at war; but they collapsed as soon as they acquired an empire. They did not know how to use the leisure which peace brought; and they had never accustomed themselves to any discipline other and better than that of war.[25]‡

* This apparent equity does not preclude offering more frequent intercourse with women to the males who demonstrate excellence in war or elsewhere.

† A brief, forward-looking note: The civic republican tradition and its various exemplars, including Machiavelli and Rousseau, did not leap over what, to their eyes, was the historic hiatus of the Middle Ages only to embrace Aristotle. Instead, they looked to the Roman Republic or to Sparta, models of civic virtue bristling with armed readiness. Aristotle was too soft by far in their eyes, diluting the bracing purity of politics with the taint of a more pacified sociality.

‡ Aristotle, citing the example of Syracuse and the tyranny of Dionysius the Elder, insists that tyrants are easily made "war-mongers, with the object of keeping their subjects constantly occupied and continually in need of a leader."[26]

The Discourse of War and Politics: From the Greeks to Today

Mature civic identity as masculine citizenship is a way of life Aristotle portrays as rich and full. Although the very existence of the *civic* as a human possibility requires constraints, its restrictions are generative, enabling human beings to communicate with one another, to organize a shared way of life, to structure moral rules. Aristotle celebrates an order that tames, and rejects the unruliness of limitless freedom in the form of individual excesses, aggressive and sexual. Some, usually the "worst sorts" in Aristotle's schema, unable to structure their individual experience in and through the constrained moral life of the *polis*, will become "out-laws," will "plunge into a passion for war."*

Once war becomes habitual, whether among Greeks or "all the un-civilized peoples which are strong enough to conquer others," "the highest honours" are paid to "military prowess; as witness the Scyth-ians, the Persians, the Thracians, and the Celts." When laws are directed "to a single object, that object is always conquest." Heartily disap-proving, Aristotle fosters instead the notion of putting war and warriors in their place. There are many ends and aims, and war itself "must therefore be regarded as only a means to peace; action as a means to leisure; and acts which are merely necessary . . . as means to acts which are good in themselves."[27]

The citizen-warrior of the *polis* is a servant of other ends. But the man "without a *polis*" is one who "at once plunges into a passion for war; he is in the position of a solitary advanced piece in a game of draughts."[28] He is decivilized, unrealized as a *polis* being. Elemental chords are plucked in Aristotle's depiction of the code of civic virtue in conflict with the disintegrative rages of the war lover. Here, too, American culture offers its own contemporary variant. Our valorized lone warrior—John Rambo of two enormously successful films, giant popular hits in the United States (but not here alone)—suggests that the inarticulate male of the deed retains his force as an emblematic presence.

Rambo is portrayed as a patriot so deeply alienated from his own society that he becomes a stateless person whose rampages are meted out variously to feckless local officials (in *First Blood, Part I* [1982]); to Russians, North Vietnamese, and CIA-bureaucratic finks (in *First*

---

* Women are necessary to, but not an integral part of, the civic schema. But domestic life has its own moral rules and purposes.

55

Blood, Part II [1985], see photo section, #6, the second and more famous Rambo film. Note that it is always "first blood," always Rambo's adversaries who draw blood first, thereby compelling him to respond. He is perpetually wronged). Rambo's loyalties are limited to other men-at-arms, veterans who have similarly suffered and bled only to be abandoned. Rambo—unsocialized, like Aristotle's man "without a *polis*"—"at once plunges into a passion for war." Should this passion overtake a whole people in the form of a structured, singular commitment to the *polis* above all and no matter what, the result is a fixation on victory, a narrative of unhappy endings for others.*

Aristotle's warnings and restrictions, picked up and played out as Christianized in the medieval West, were central to alternatives to the genealogy of armed civic virtue. Plato's ideal of unity and his celebrations of the Guardian/warrior nexus, together with constructions of the Greek and Roman political experiences, reappropriated by the so-called civic republicans, bring the story of armed civic virtue into early modern European history.

The Ideal Republic:
Machiavelli and Rousseau

Machiavelli, celebrated as the author of *The Prince* (1532) and *The Discourses* (1531), also wrote *The Art of War* (1521).[29] For Machiavelli, even peace is warlike, for politics is a constant struggle for power. His characterizations of political life bristle with the metaphors of war and posit, as the penultimate goal of that life, victory, success, the grabbing and holding of power. Machiavelli's *virtù*, a particular understanding of political virtue, is linked to reigning notions of masculine virility. Over three centuries later, this nexus is reiterated in Alexis de Tocqueville's discussion of the Roman republic in *Democracy in America*. Citing Plutarch's *Life of Coriolanus* to the effect that "martial

---

* Perhaps demonstrating once again the way we play footnote to the Greeks, the matter of commitments, public and private, and teleologies with winners and losers, figure in current politics of war and peace among feminists and nonfeminists alike—matters that I treat in chapter 7.

56

prowess was much more honored and prized in Rome than all the other virtues, in so much that it was called *virtus*, the name of virtue itself, by applying the name of the kind to this particular species: so that *virtue* in Latin was as much as to say *valor*," Tocqueville opines: "Can anyone fail to recognize the peculiar want of that singular community which was formed for the conquest of the world?"[30]

In the introduction to his *Discourses*, Machiavelli proclaims that his own age—the sixteenth century in Italy—is a nadir because it has forgotten the "prodigies of virtue and wisdom displayed by the kings, captains, citizens, and legislators who have sacrificed themselves for their country." Ignoring "the history of ancient kingdoms and republics," his contemporaries show not the "least trace" of their ancient virtues.[31]* He holds up, in *The Prince*, as exemplary the canny ruthlessness of Cesare Borgia who seduced his opponents with promises and then stabbed them in the back. The appropriate moral test to apply to Borgia's successful deceits and stealthy murders is whether they worked *from the standpoint of statecraft*, not from the perspective of Christian morality. Within a picture framed by *raison d'état*, Borgia was successful; thus, Machiavelli places him high in the pantheon of rogue heroes.

Machiavelli's preference, however, is not for tyrants but for the armed and virtuous republics of the ancient world. The first duty of Machiavelli's prince is that he be a soldier and create an army of citizens prepared to defend and to die for the *respublica*. In a passage from the *Discourses*, his passionate love for his *civitas* soars and takes flight: "When it is a question of saving the Fatherland, one should not stop for a moment to consider whether something is lawful or unlawful, gentle or cruel, laudable or shameful; but putting aside every other consideration, one ought to follow out to the end whatever resolve will save the life of the state and preserve its freedom."[33] His goal is *self-sufficiency*; his ideal, that of a polity akin to a singular armed body. The citizen = the self-sufficient, armed warrior = the armed militia = armed civic virtue/the popular state: this is the Machiavellian recipe for civic autonomy.†

---

* Jeff Weintraub has identified the Roman republic as the central myth of the civic republican tradition.[32] Machiavelli reappears later in this chapter as an exemplar of the constituted tradition of realism, the reigning paradigm of IR (international relations).

† Writes J. G. A. Pocock in his classic *The Machiavellian Moment*: "There must be the political

The Machiavellian citizen, like the Platonic Guardian, must dedicate his life and actions to the republic or the city; the civic good is paramount. But where Plato requires knowledge of the transcendent Forms as the claim to legitimacy of his ruling élite, Machiavelli substitutes *will*. Each—Guardian and armed citizen—is subjected to, and freely subjects himself (only men can be citizens for Machiavelli)—or himself and herself, in the case of Plato's Guardians—to a rigorous discipline of mind and body aimed at the unity and autonomy of the *polis*. And that good is predefined by Plato and Machiavelli as the forging of one out of the hotchpotch of many. Plato's architectonic scheme eliminates internal political struggle; once in place, his city is set up to run smoothly, without politics. Machiavelli sees struggle as endemic and would create citizens fit for struggle so that they can be mobilized as effectively as possible in any situation, "hot" or "cold." *Si vis pacem, para bellum*—or, roughly, "to have peace, you must prepare for war."

Machiavelli's story of militarized citizenship is a narrative revolving around a public-private split in and through which women are constituted either as "mirrors" to male war making (a kind of civic cheerleader) or as a collective Other, embodying the softer values and virtues out of place within, and subversive of, *realpolitik*. Immunized from political action (save machinations in the boudoir or behind-the-scenes stealth in behalf of husband or lover), Machiavelli's female may honor the penates but cannot embark on a project to bring alternative values to bear on the civic life of society. Pocock calls Machiavelli's "militarization of citizenship" a potent legacy that subverts consideration of alternatives that do not bind civic and martial virtue together.[35]

If military preparedness is the *sine qua non* of a virtuous polity, and women, in this narrative, cannot embody such armed civic virtue—a task for the men—women are nonetheless drawn into the picture: as occasions for war; as goads to action; as designated weepers over the tragedies war trails in its wake; or, in our own time, as male surrogates mobilized to meet manpower needs for the armed forces.*

---

conditions which permit the arming of all citizens, the moral conditions in which all are willing to fight for the republic and the economic conditions ... which give the warrior a home and occupation outside the camp and prevent his becoming a ... mercenary whose sword is at the command of some powerful individual."[34]

* The judgment implied here is harsh, with the modern female soldier constructed as a male

Although Machiavelli's occasional references to women are of little major theoretical importance, his discourse does feature Fortuna, a bitch-goddess who controls about one half of all human affairs, and whom he compares to an impetuous river or a fickle woman who must be mastered and conquered by force; he proffers an assessment of her in "How States Are Ruined on Account of Women."[36] Machiavelli has no extended discussion of the relation of family (women's world) to polity (men's world), but passages from this chapter of *The Discourses* are instructive insofar as they hold women responsible for the damage done when men seek revenge of women's honor.[37]*

Through it all, Machiavelli's emphasis on war as vital social force as well as dire necessity, as creator of social solidarity as well as destructive tragedy, comes through loud and clear. As the penultimate form of collective struggle, war symbolizes what solidarity, one for all and all for one, is about. Civic virtue is armed and willful, the source of legitimacy, stability, the basis of the *respublica*. Machiavellian themes echo throughout the subsequent history of political discourse in the West.

But the echoes go beyond texts, sounding at this moment in the techno-dreams of modern American makers of the machines of war as they appeal to the policy elite in government responsible for putting out defense contracts. An advertisement in *Air Force*, placed by McDonnell Douglas to sell its F-15 Eagle fighter, declares in large, bold print:

EYES ON THE OLIVE BRANCH,
BUT ARROWS AT THE READY.

The ad writers have a finger on our own variants (or deformations) of the Machiavellian moment as they proclaim, quoting the Father of Our Country: "To be prepared for war is one of the most effectual means of preserving peace." The F-15 Eagle is "a manifestation of the

---

in drag. From the standpoint of the state, this is what she is—a militarized, masculinized female. Her own perspective will be taken into account, as counterpoint, in chapters 5 and 7.

* Augustine, in *The City of God*, deconstructed the Roman ideal of "woman's honor." He discussed multiple brutalities: against Roman women by invading forces; against the ravishers of Roman women by Roman soldiers; and women against themselves, given the ideal of honor, once having been ravished.[38] Resurrecting honor, Machiavelli must reconstruct the noble statues Augustine had taken a sledgehammer to.

Great Seal's symbology. Strong enough to win, awesome enough to deter. By its very presence *it is an expression of the national will.*" The beat goes on.

When Jean-Jacques Rousseau, writing more than two centuries after Machiavelli's death (1527), proclaims that "true Christians are made to be slaves" and ill suited to citizenship, he, too, echoes Machiavelli. The terms "Christian" and "republic," Rousseau insists, "are mutually exclusive. Christianity preaches nothing but servitude and dependence." Worst of all is Roman Catholicism, a religion "so manifestly bad that it is a waste of time to amuse oneself by proving it.... All institutions that put man in contradiction with himself are worthless."[39] The plangent note sounds: the polity must be as one; the national will must not be divided; citizens must be prepared to defend civic autonomy through force of arms; whatever puts the individual at odds with himself is a threat to "*la nation une et indivisible.*" The body individual and the body politic must be driven by a single motor. Hence Rousseau's hostility to any save a watered-down, deistic civic religion constructed as a prop of rather than a possible irritant to the body politic.*

Rousseau's animus is meted out to any *particular interest* that might block the general interest or will, driving a wedge between the citizen and the wider social body. The *citizen* is not to be confused with the bourgeois, a man given over to private interest and commerce. The transition from private to public person is ritualized by Rousseau as a rite of passage analogous to religious conversion, on the one hand; to marriage, on the other. Mimetically, the pre-citizen puts "his will, his goods, his force, and his person in common ... and in a body we all receive each member as an inalienable part of the whole." We strip off the old and put on the new person, not in Christ, but in the body politic.[40] Citizenship is constituted in the interstices of the self, inscribed in the flesh, felt in the bones.

A true city, or *cives*, emerges not out of a collection of houses clustered on the same site, but "citizens make the City." And the citizen must give his all; must be prepared to fight. Either a citizen or a "debased slave": the choice, Rousseau in effect proclaims, is ours; and the fault, if we are slaves, lies not in the stars but in ourselves. That the

* Rousseau wasn't fond of atheists either, fearing they might be feckless, unwilling to commit themselves.

citizen and the state are one public body, but the private body and the state do not form such an identity, is clear in Rousseau's discussion of war:

> War is not a relation between men, but between powers.... The end of war is the destruction of the enemy State. One has the right to kill its defenders as long as they are armed, but as soon as they lay down their arms and surrender, they cease to be enemies, or rather instruments of the enemy, and one no longer has a right to their lives. One can kill the State without killing a single one of its members.[41]

The apparently enigmatic last sentence of this passage is not so mysterious. The state can be killed through an annexation without armed struggle, for example, or through a pre-emptive attack that renders the enemy defenseless—stateless, its citizens de-civilized—before they have a chance to fly to arms. Footnoting Machiavelli throughout the *Social Contract* (1762), Rousseau's image of robust virtue is of a city on a hill, shining with civic purity, and a city surrounded by walls, a fortress that all citizens will defend and from which none but the faithless will flee in time of dire testing.

Celebrating military virtue, reappropriating Machiavelli's virilized discourse, Rousseau rails against softness and decadence. Military virtue "died out" among the Romans the more "they became connoisseurs of paintings, engravings, jeweled vessels and began to cultivate the fine arts." The ancient Greek republics, "with that wisdom which shone through most of their institutions, forbade their citizen the practice of those tranquil and sedentary occupations which, by weighing down and corrupting the body, soon enervate the vigor of the soul."[42] Rousseau advises the Poles to "look with a tolerant eye on military display. ... But let all kinds of womanish adornment be held in contempt. And if you cannot bring women themselves to renounce it, let them at least be taught to disapprove of it, and view it with disdain in men."[43]*

* Although she would have disdained Rousseau's use of "womanish" to characterize the collective male penchant for self-advertising adornment, Virginia Woolf would otherwise have concurred with Rousseau's censure of the overly refined display of the male person with fancy uniforms, excessive medals, and so on. On all other matters, they part company for Woolf derealizes the citizen (see chapter 7).

Civic education must turn boys into men, keep them busy, mold them to duty. Sparta as refracted through Plutarch is proffered as exemplary, "the excellent regulations of Lycurgus [Spartan ruler and lawgiver]" termed "in truth monstrously perfect" because they set forth the disciplinary techniques required to turn children into "noble youths." Rousseau also honors Spartan mothers whose "Sayings" Plutarch detailed in volume III of his *Moralia*, reproducing tales, anecdotes, and epigrams that constructed the Spartan woman as a mother who reared her sons to be sacrificed on the altar of civic need. Such a martial mother was pleased to hear that her son died "in a manner worthy of [her]self, his country, and his ancestors than if he had lived for all time a coward." Sons who failed to measure up were reviled. One woman, whose son was the sole survivor of a disastrous battle, killed him with a tile, the appropriate punishment for his obvious cowardness. Spartan women shook off expressions of sympathy in words that bespeak an unshakable civic identity. Plutarch recounts a woman, as she buried her son, telling a would-be sympathizer, that she had had "good luck," not bad: "I bore him that he might die for Sparta, and this is the very thing that has come to pass for me."[44]

One of the great prophets of armed civic virtue, Rousseau's ideas were absorbed by leading thinkers and activists of the French Revolution, most importantly Robespierre. As translated into the politics of the revolution, military duty and martial motherhood is one of the refracted outcomes of Rousseau's teaching. A second is a preoccupation with civic virtue which, in the hands of men who claimed Rousseau as inspiration, got realized as political terror.*

A demanding vision of total civic virtue requires the presence of "enemies within," traitors whom true citizens must constantly be on guard against—as Robespierre's rhetoric makes clear. In his brief speech during the trial of Louis XVI, Robespierre appealed to "vigor" no less than six times and to "zeal" nearly as often. The "republican virtues," he feared, would be weakened by clemency, by natural feelings and

---

* Hannah Arendt's criticisms are sharp: "Robespierre carried the conflicts of the soul, Rousseau's *âme déchirée*, into politics where they became murderous," for they necessitated a search for "pure motives," a ferreting out of domestic enemies, a scouring of the body politic for possible traitors who wore what Robespierre called the "mask" of virtue and patriotism. It is the war upon hypocrisy that transformed Robespierre's dictatorship into the Reign of Terror, and the "outstanding characteristic of this period was the self-purging of the rulers."[45]

sensibilities that must be suppressed if justice was to be done. Thumping clemency, he deflected it as a human virtue by insisting that it is "barbarous" if it consorts with "tyranny," and that was precisely what those who found themselves sympathetic to Louis were guilty of. The words "pretext," "art," "mask," "intrigue," "plot," "hypocrisy" pepper the text as "our enemies" without and within, "perfidious" and "reptiles," erode the zeal and ardor of true republican virtue "which is always in the minority on this earth."[46] Armed republican civic virtue, constituted to protect, define, and defend a way of life, translated into statist societies, is here unleashed as nationalism, and the dream of discipline is "made national."[47]

Citizen Carnot, the member of the Committee of Public Safety entrusted with carrying through on universal conscription, explained: "Henceforth, the republic is a great city in a state of siege: France must become one vast camp and Paris its arsenal."[48] The law itself conscripted "every Frenchman," including "young men," "married men," "women," "children," "old men." The third article of the 1789 Declaration of the Rights of Man, in a celebration of collective epiphany, held that "the Nation is essentially the source of all sovereignty; nor can any individual or any body of men be entitled to any authority that is not expressly derived from it"; and chapter 4, to punctuate the point, added, "The law is an expression of the will of the community."[49]

RESISTANCE TO THE REPUBLIC OF ARMED VIRTUE

That this "will" could be brought to bear successfully on any recalcitrant wills becomes clear as one considers the violent and poignant wrenchings of compulsory military service, the chief "process which drew men out from their regional *pays* and compelled them to be conscious of belonging to a wider community." Impressment for military service, for defense of one's locale, was nothing new under the sun but; as the historians Michael Howard and Eugene Weber and the memoirist Pierre-Jakez Hélias recount, the power to draw hundreds, then thousands, of men out of their previous way of life was unprecedented prior to 1793 and the *levée en masse*. Its enforcement, and that of subsequent conscription ordinances, brought into being an organization "serving a new, vast entity" as "the men of Normandy,

Languedoc, Picardy, Franche-Comté, Auvergne, Limousin, Provence, were hauled away and turned into *Frenchmen*. It was a process so disagreeable that one province, Brittany, fought a bitter civil war to avoid it."[50]

Peasant communities resisted longest and hardest. Finally, however, they were suborned through a combination of "carrot-and-stick" stratagems deployed by the central and centralizing forces. Men fought to avoid military service by buying substitutes—if they could afford it—until substitution was abolished in 1873 under the Third Republic, which also established a five-year term of military service. Eugene Weber adds that, at the time this reform was introduced, the government also initiated "a wealth of dispensations, mostly for the educated classes, ranging from outright exemptions to a 'voluntary' one year tour of duty on payment of a 1500 franc fee."[51] In 1889, the term of service was reduced to three years, previous exemptions were dispensed with, and the move toward true universal conscription, as a dream of national discipline—with the army as the central agency for creating Frenchmen by "civilizing" provincials—came closer to realization.

The story of conscription in the French Republic is a story of sporadic and finally failed resistance. Official records turn up tale after tale of draft evasion or desertion. Weber cites excerpts from reports of disconcerted central government observers: "Pronounced antipathy for military service. . . . Very reluctant to join the army. . . . Great tendency to avoid service by self-mutilation, decamping, or attempted bribes. . . . Profound horror for military service." One small Bourbonnais village near Lavoine devised an even more ingenious method of evasion: "almost all boys at birth were declared as girls," a practice Weber finds still being used in the 1870s.[52] His conclusion is that the anti-militarism of the twentieth century, often presented as a story of "red propaganda," actually has its roots in the frightened, desperate manipulations of peasant peoples who felt no sense of national identity but were compelled to give their fathers and sons over to its service, to the extent of forefeiting life itself.

The army remained "theirs," not "ours," until toward the end of the nineteenth century, after the Franco-Prussian War, when the army as a "school of the fatherland" finally worked its will by wrenching young men into another language, stripping them of their local *patois*, shifting their allegiance, finally, to the national flag through a series of quasi-religious ceremonies and outright bribes in the form of "Everyday, meat and soup/Without working, without working./Everyday, meat and soup/Without working in the army." By raising the standard of living in the army above those of much of the rural classes, many of the harsher, coercive measures could be alleviated as the dream of soup and "white bread" took over.*

Pierre-Jakez Hélias, in his story of life in a Breton village, a monument to a peasant culture now dead in a society homogenized by universal military service, public education, and mass culture, evokes the anger and alienation war—the Great War, in this case—brought to his own village. Rumors of war had reached the village, but life went on. His mother and father were "toiling in the Meot field, cutting the wheat with sickles." But late in the afternoon,

> the bells of the parish church were set to ringing in a mode that made one think the sexton had lost his head. Actually the poor devil was signaling a fire that was to last for over four years throughout the world. How could he have found the right tone? He was going from one bell to another, striking them with the awkwardness of despair. But everyone clearly understood his extraordinary language.
>
> My father picked up his sickle and wielded it a few times, but more and more slowly. Then he bent one knee to the ground and lowered his head. Suddenly he stood straight up, threw his tool far into the distance, and started to walk toward the town, through the fields, without ever once unclenching his jaws. My mother sat on the ground and wept into her apron.[53]

Hélias didn't see his father again until the end of the war, but he heard strange words, *French* words for which there was no Breton equivalent—words like "the front," "the krauts," "trenches," "shells."

---

* In the Austro-Hungarian empire—which faced the even more daunting task of trying to forge an army out of disparate national entities many of which hated each other, the empire, or both—young men from its core and its periphery, from Vienna and from the outskirts of Galicia, all became "sons of His Majesty" fused to the emperor through a holy indissoluble bond. The effort here was to engraft essentially feudal bonds of loyalty, a pledging of one's troth through a ceremony of *adoubement*, blessed by the clergy, onto a multinational, multilinguistic mix of men. Just as the medieval feudal bond was a factor aimed, in part, at preventing the breakdown of the territorial principality, so its use into the twentieth century aimed to preclude the "Balkanization" of large, constructed political unities.

Not having words in their own language to convey the meaning of the war that had taken so many of their men, the peasants of his village suffered *from* the effects of the war but felt estrangement from it. Then one day, he reports, "we learned yet another French word ... which echoed from mouth to mouth like applause: *armistice*." Soldiers and sailors returned, or at least some of them. Many were dead, and some were seriously disabled lying in hospitals far away. But others found their worlds broadened and stayed on elsewhere, seeking jobs and wives in other areas. The soldiers who returned had picked up some new ideas, but so had their wives who "found it difficult to give up" the prerogatives they had assumed with the menfolk away.[54]

"Liberty, Equality, Fraternity"—that was what it was supposed to be about; and on Armistice Day, Hélias heard these words and watched a monument being built. But the Republic was not yet "real," not yet constitutive of his own identity, not until schools took over and, forbidden to speak his own language, he, too, became "French." Their peasant identities having become suspect, humiliated by the epithet *la vache* ("cow," a derogatory term for a peasant), repudiating the *mother* tongue, he, and others, became Children of the Republic—they along with the men who had "saved France," who had "beaten the Krauts," who had learned to sing *La Marseillaise*.

Caught up in the grips of *la patrie*, the indivisible, binding, indissoluble female fatherland, "France" finally comes into being. But it is a rather recent historic development, one of this century.

WOMEN ON ROUSSEAU'S DISCOURSE: MARY WOLLSTONECRAFT

The line of descent I have traced takes us from Rousseau's Geneva, an ideal city-republic surrounded by *real* walls, to the nation-state encompassed by no such palpable barrier, fortified instead by mytholegalistic grandeur. Borders are militarily and legally established that others dare not transgress, just as to breach the walls of the earlier, fortified city was an act of war. The modern nation-state, no longer an encased *civitas*, cannot lock its wandering alienates out (as young Rousseau found himself locked outside the gates of Geneva), at least not literally—but it can do so *literarily*, with the use of documentation that permits or denies entrance and egress along with police and soldiers to shore up these barriers and boundaries. The *fortifications* are rebuilt, then, symbolically, legally, through the vast pretensions of sovereignty—all signifying the extraordinary shift from the local *patria*, the original site of patriotism, to an inclusive nationalism that must *defeat* local identities or locate them as subordinate and inferior. This is the central plot as the story of armed civic virtue unfolds. But there is a subtext in this narrative, a debate about the role of women in politics, a conflict of self/other definition, a collision that becomes more and more pivotal.

Rousseau is a key player in this tale, too, for, try as he might, he could not put this matter to rest, could not quell the tumult of decisive encounters. This apparent inconsistency speaks instead to his qualities as an articulator of paradox and conundrums and, as well, to transformations in the sensibilities and identities of men and women as historic subjects. Rousseau's representations of women strain at the constraints demanded by his vision of civic good and of feminine sensibility, a dense interweaving of public and private. Unintentionally, he helps frame, for the first time, feminist discussion as a self-conscious historic phenomenon. The generative force of his discourse, already limned in the actions of Robespierre, takes a different turn in the writings of a woman who was one of his great admirers—and critics.

Mobilized historically with the *levée en masse* (the conscription of the entire French population for war), women had undergone a prior mobilization of a rather different sort discursively. The Revolutionary conscription law called for women to "make tents and clothing, and work in hospitals," and for children to "turn old linen into bandages." The assignment of such tasks assumes a prior commitment to the nation, to its survival, and to its victory over enemies without and within. How does such loyalty come into play in the first place? If women are not fully *citizens*, through what social relations and symbolic representations, through what webs of semiotically charged events and actions, are they entangled with the life of the body politic?

For Rousseau the answer lies in the inner relations between private and public virtue. In his *Letter to M. D'Alembert on the Theatre*, Rousseau describes the "cackle" of women's societies, noting that much of their conversation revolves around the behavior of other women. He approves of this, arguing that gossip gives women a censoring role.

By assuring that a deviant will be talked about, sometimes mercilessly, or even ostracized, women can use their collective speech to prevent scandal and to promote virtue. "How many public scandals are prevented for fear of these severe observers?" he writes.[55] As arbiters of private or local morality, women are in a powerful position to keep others in line, to prevent the eruptions of *individualities* corrosive of the collective.*

But village gossip is icing on the cake of women's virtue. Rousseau has a more complex story to tell, a tale of love of fatherland and mother's milk. Virtuous polities require virtuous families. Children must imbibe love for the fatherland (*la patrie*) with their mother's milk.[56]† Romantic love between men and women sets the basis for the civic family, the family that inculcates a passion for political virtue (although it also may threaten that virtue if it is *too* insular and absorbing). Even as the coming together of male and female results in a mutually interdependent small and particular society, so Rousseau's vision of a body politic, captured as I noted earlier in the metaphor of the "general will," constitutes a potent moment of collective identity.

Mother and mother's milk serves as a foundation for civic-spiritedness. Thus Rousseau in *Emile*:

> The first education is the most important, and this first education belongs incontestably to women; if the Author of nature had wanted it to belong to men, He would have given them milk with which to nurse the children. ... The laws—always so occupied with property and so little with persons, because their object is peace not virtue—do not give enough authority to mothers. However, their status is more certain than that of fathers; their duties are more painful; their cares are more important for the good order of the family; generally they are more attached to the children. There are occasions on which a son who lacks respect for his father can in some way

---

* Andrzej Wajda's 1984 film *A Love in Germany* portrays the shift from "gossip" to "informing" as he details the ways in which the ordinary activities of provincial German village women acquire extraordinary power to compel and to punish, given certain shifts in the larger public world over which women exercise no effective control. In light of this shift, and given the fact that many of the women are ambitious *through* their husbands and fearful *for* themselves should something happen to their husbands, a tragedy of wartime takes shape. Gossip is no longer just about a woman who is breaking the sexual code but becomes an act of informing to the authorities *on* a woman who transgresses the racial/sexual laws of the Nazi state.

† As noted earlier, *la patrie*, or "fatherland," is feminine, requiring the feminine definite article *la* rather than the masculine *le*.

be excused. But if on any occasion whatsoever a child were unnatural enough to lack respect for his mother—for her who carried him in her womb, who nursed him with her milk, who for years forgot herself in favor of caring for him alone—one should hasten to strangle this wretch as a monster unworthy of seeing the light of day.[57]

Similarly, Rousseau finds the citizen in whom love of the fatherland as a civic *mother* does not beat steadily and true, to be a monster, an unworthy wretch. Just as treason to the mother warrants strangulation, treason to the state calls for draconian punishment, public shaming, and execution. Just as the mother's watchful eyes must be ever on her developing child, so the

citizen shall feel the eyes of his fellow-countrymen upon him every moment of the day; that no man shall move upward and win success except by public approbation; that every post and employment shall be filled in accordance with the nation's wishes; and that everyone—from the least of the nobles, or even the least of the peasants, up to the king himself, if that were possible—shall be so *completely dependent upon public esteem as to be unable to do anything, acquire anything, or achieve anything without it.*[58] (Italics added.)

Mothers figure centrally in this dream of civic unity, but not *as* citizens; rather as mothers of citizens-to-be and of mothers-to-be of citizens. Enter Mary Wollstonecraft, feminist foremother, in one of the first of many feminist skirmishes with Jean-Jacques. Wollstonecraft's *A Vindication of the Rights of Woman* was published in the first edition in 1792, three decades after the publication of both *Emile* and the *Social Contract*. Rousseau had been dead but fourteen years and the French Revolution had erupted three years earlier; indeed, revolutionary ferment was at its height throughout 1791 when Wollstonecraft was writing her tract.

Seeking to extend the French Revolutionary's proclamation of the rights of man, Wollstonecraft insists on equality without regard to sex. Although she does not direct her thoughts toward them—for this would confirm rather than put pressure upon Rousseau's notions—presiding revolutionary images at the time included paintings of the militant Spartan mother, which began to appear in the 1790s, and

Liberty as represented by Marianne, a young female militant with one breast bared.*

While Wollstonecraft's special nemesis is Edmund Burke of *Reflections on the Revolution in France*, Rousseau is an admired adversary and mentor whose teachings were, to her mind, marred by his unacceptable views on men and women, sexual identity, and citizenship. Wollstonecraft insists that women must be active citizens if they are to pass civic virtue on to their young. Rousseau, however, precludes direct female political participation in part because of the taut nexus he draws between being a citizen and bearing arms, a sex-linked and immutable civic difference to his mind.

Men alone, for civic republicans like Machiavelli and Rousseau, have the bodies of defenders; they alone can serve as soldiers. Wollstonecraft acknowledges that "bodily strength seems to give man a natural superiority over woman; and this is the only solid basis on which the superiority of the sex can be built." She effaces this distinction, however, by calling for identical educations, hence knowledge, for the two sexes so that as moral and rational creatures they will have the same virtues.[60]

Women, absorbed with caring for their vulnerable infants (Rousseau), or caught up in sexual intrigue (Machiavelli, or "fallen" women for Rousseau), cannot go forth to defend the polity through force of arms. Rousseau, in a chilling passage from *Emile*, describes the female "citizen" as follows: "A Spartan woman had five sons in the army and was awaiting news of the battle. A Helot arrives; trembling, she asks him for news. 'Your five sons were killed.' 'Base slave, did I ask you that?' 'We won the victory.' The mother runs to the temple and gives thanks to the gods. This is the female citizen."[61]† The potent love of

---

* The civic republican world, for all its female icons, has been strongly male-dominant. Women voted for the first time in a parliamentary election in France in 1945. Although not a citizen, the collective "she" was to be a source of civic virtue. There is a certain contrast to the Church, in which women could be elevated to beatitude or sainthood and had a place in religious orders and educational and charitable organizations. George Steiner notes that French Revolutionary women reinscribed, for their own purposes, representations of the Lacedaemonian women, companions in arms to their husbands and the "matrons of Republican Rome, Brutus, and Cato's equals.... Certain women—Mme Roland, Charlotte Corday—performed heroically and sacrificially during the French Revolution. They referred themselves to Plutarch ('Cato's daughter') rather than to the anarchic solitude of revolt in Antigone."[59]

† Rousseau got this story from Plutarch, whose Spartan mother tells the "vile varlet": "I accept gladly also the death of my sons."[62]

mother country, and willingness to serve and protect her, will shrivel on the civic vine if mothers no longer figure overpoweringly in the affections and upbringing of their children.

Wollstonecraft will have none of this: she wants the virtue without the qualification "armed" and would excise the martial dimensions from Rousseau's vision. Holding that women, too, would do their part should the polity be attacked, she shares a vision of civic motherhood. But Rousseau, in her view, celebrates not virtue but "barbarism." He valorizes the Romans who conquered and destroyed but did not extend "the reign of virtue." He "exalts those to demi-gods who were scarcely human—the brutal Spartans, who, in defiance of justice and gratitude, sacrificed, in cold blood, the slaves who had shewn themselves heroes to rescue their oppressors."[63]*

Condensed, Wollstonecraft's argument comes to this: In his rush to rescue virtue and deflect vice, Rousseau misidentified virtue. He missed the truly "gigantic mischief" of arbitrary power and "hereditary distinctions" that, together with a "standing army," are "incompatible with freedom." "Subordination" is the chief sinew of "military discipline," and "despotism is necessary to give vigour" to the military enterprise. Notions of honor may be apt for the commanding few, but the vast majority of soldiers are a mass swept along by coercion and command. How can *this* serve as a model of civic probity? With her reply to Rousseau, and her certain conviction that if "women share the rights" they "will emulate the virtues of man" and "grow more perfect when emancipated,"[64] Wollstonecraft set the terms of discourse for liberal queasiness over civic republicanism and female ire over "male tyranny."† But she continued to endorse many of the ends and aims of civic republicanism: civic virtue and autonomy, shorn of its martial dimension, most importantly.

If Rousseau's arguments locate him in a paradox—that women, untrained for civic identity must nonetheless transmit, to their sons, a passion for that identity and, to their daughters, a passion for the exemplary families necessary to nurture that identity—Wollstonecraft

---

* Also, Wollstonecraft rejects Rousseau's insistence that man is not meant to be a carnivore, a killer and eater of other creatures; for her, we are naturally meat eaters just as we are naturally more sociable than Rousseau makes provision for.

† I take up war and liberalism more fully in chapter 7.

runs ashoal as well. She *assumes* a national identity, paying no attention to the violent manner in which the French Revolutionaries molded that identity by yanking young men out of their local identities, disciplining their bodies to armed purposes, scraping off the insignia of their particularity by the visible sign of putting diverse human elements into identical uniforms. The uniform as a sign betrays nothing of its wearer's "origins": that is part of its intent.* In her narrative, Wollstonecraft attempts to demobilize the male virtues into which women are to be educated. But might not her endorsement of the locution *Rights of Man* in itself incorporate tacitly the excesses Wollstonecraft tries to defeat?

The Rights of Man, as Hannah Arendt points out, is a generic concept shorn of historic specificity.[66] These rights, celebrated by Wollstonecraft, are located as the foundation of a new body politic: man as natural being = man as civic being. If the state of nature is a metaphorical construct, natural human rights functions metonymically. The dissolving of the civic into the "natural" man effects a powerful substitution: man as genus *Homo* acquires foundational status in that his rights ground the new body politic. The natural man is politicized from the day of his birth as a being already and entirely marked by the social order, destined to serve as its beginning and its end.

The people, a potent rhetorical crystallization, takes on teleological status: it is that which must and will come into being once autocratic or self-interested barriers and brakes are removed. Space for withdrawal or for dissent from the civic virtue of the mobilized people shrinks as the natural man absorbs social and political man.† Wollstonecraft's "rights of woman" are the rights of man rendered as sex blind or gender inclusive. If the "standing army" is a vice, as Wollstonecraft insists, it is a vice brought into being by the constellation of forces and ideas she herself endorses. The army model of discipline and mass force—in contrast to the medieval feudal vision of coteries of armed

---

* In March 1986, in a five-to-four decision, the United States Supreme Court ruled that Jewish members of the armed forces had no right to wear yarmulkes, or skullcaps, as a sign of identity and piety: the basis for the decision was the military's need to "foster instinctive obedience, unity, commitment and *esprit de corps*." This *group* identity must take precedence over other signs *save those of rank*.[65]

† These remarks owe much to Arendt's discussion, but I view the matter from a slightly different angle, given my concern with what happens to human bodies in different social orders.

men at war, individually distinguished by banner, insignia, weapons, horse draperies—is standing *in situ* when it doesn't stand in fact: it is a possibility that can be brought into being given the structure of the centralized nation-state.

These remarks are not intended to deflect from Wollstonecraft's critique so much as to locate it inside a cluster of problems that engaged both her and Rousseau, and that burst onto the stage of history in dramatic form, shaping subsequent political discourse and events, with the upheavals of the French Revolution, including the triumph of the state ideal and its embrace by philosophers and military men outside of France.

The Nation-State

HEGEL'S VISION OF THE STATE

The nation having been mobilized—that is a "nation" created out of a congeries of competing local identities—the triumphant nation-state, or *kriegstaat* (literally, "warfare state"), erases alternative notions of civic being, or, at least, shoves them into the background. Born in 1770, the German philosopher Hegel as a young man celebrated the idea of the French Revolution. Just as the French Revolutionaries freed men from the fetters of king and priest, so philosophy would free the mind from the chains of narrow religiosity and cramped parochialism. Although critical of Robespierre, and of terror without limits, Hegel shared the lofty aims of folding the particular bits of a scattered nationality into the holding bin of the state. He would re-create the "we" of the classical republic in and through a series of complex mediations. While direct and immediate "we"-ness is not possible in large domains, with their multiplicities of human associations and identities, we must somehow—preferably not through terror—reach the apogee, the state.[67]

The divisions of bourgeois society are a threat: a threat to freedom, to self-conscious human subjects, to ethical life itself as embodied in its three central loci—the family, civil society, and the state. The French

73

Revolution had inaugurated a "new age" for Hegel and his contemporaries. It was a world complicatedly at odds with itself in ways that contained possibilities for resolution, he opined. For example: we require families. Without the relations of intimate domesticity, life would truly be Hobbes's *bellum omnia in omnis*, "war of all against all"—a nightmare. But families ensure human division into independent entities "presided over by womankind." For civilization and universal reason to triumph, the family must be interfered with and the individual *self*-consciousness it engenders dissolved. Woman, for Hegel, is essential yet an "internal enemy"—a view similar to points in Rousseau. That, in fact, is the irony of woman's situation in relation to the state (hence to war) as Hegel sketches it.[68]

Hegel's is a grand vision of the state—the "actuality of an ethical idea," in his characteristic language. As a state-identified being, the *self* of the male citizen is fully unfolded, made complete. Because his freedom is dependent upon that of others, he must get beyond the individualistic freedom of bourgeois civil society. Enter the state and a way, first, to meet the human needs left unsatisfied by the competitive individualism of civil society and, second, to transcend the conflict endemic to that sphere. The state is the arena that calls upon and sustains the individual's commitment to universal ethical life, satisfying expansive yearnings through the opportunity to sacrifice in "behalf of the individuality of the state."[69] For with the state comes not simply the possibility but the inevitability of war.*

War transcends material values. The individual reaches for a common end. War-constituted solidarity is immanent within the state form. But the state, hence the nation, comes fully to life only with war. Peace poses the specific danger of sanctioning the view that the atomized world of civil society is absolute.[71] In war, however, the state as a collective being is tested, and the citizen comes to recognize the

---

* The human work of courage plays itself out in "the genuine, absolute, final end, the sovereignty of the state." The modern world guarantees this end by transforming personal bravery into something impersonal, for "thought has invented the gun, and the invention of this weapon, which has changed the purely personal form of bravery into a more abstract one, is no accident."[70] Should the reader find this passage somewhat obscure, he or she is not alone. A few brief words of explanation: For Hegel, History has agentic force, acting upon and through not just individuals but institutions, complex relations, and so on. History has a teleology writ large: it is going someplace, and Hegel knows where. So does Marx, of course, though their engines travel down somewhat different tracks. But that is another story.

state as the source of all rights. Just as the individual emerges to self-conscious identity only through a struggle, so each state must struggle to attain recognition. The state's proclamation of its sovereignty is not enough: that sovereignty must be recognized. War is the means to attain recognition, to pass, in a sense, the definitive test of political manhood. That state is free that can defend itself, gain the recognition of others, and shore up an acknowledged identity. The freedom of individuals and states is not given as such but must be achieved through conflict. It is in war that the strength of the state is tested, and only through that test can it be shown whether individuals can overcome selfishness and are prepared to work for the whole and to sacrifice in service to the more inclusive good.

Finally, for Hegel, war is a reminder of the finiteness of individual existence. The awesome power of negativity, of death writ large, makes itself felt as citizens are drawn out of themselves into a larger purpose. Of course, men and women are drawn out differently. The woman gives up her son. The man becomes what he in some sense is *meant* to be by being absorbed in the larger stream of life: war and the state. In all this, Hegel does not so much glorify war as valorize History. War is a necessity that has played and will always play a vital historic purpose. Hegel's picture of war is of conflict limited in scope for the most part, only occasionally requiring full mobilization. But whether limited or total, war absorbs a nation and its individual players more fully in a communal, not simply an individualistic, freedom. Life seems larger, somehow, and even more precious because it is threatened.*

To those readers who are not professional political theorists, these brief passages no doubt make Hegel seem an obscure and perhaps dyspeptic thinker. But he remains essential in any consideration of women and war, of men in war. Just as the Machiavellian moment seems never to have been quite overcome, neither has the Hegelian spirit evaporated in the thinner air of twentieth-century pragmatism. We still have trouble accounting for modern state worship. The mounds of bodies of combatants and noncombatants alike sacrificed to the conflicts of nation-states compel us to take seriously Hegel's story of identity, personal and collective, through conflict.

* Hegel does warn that protracted war may backfire, may have debilitating effects, as does any war that is fueled by rapaciousness rather than by a struggle for recognition.

ARMED CIVIC VIRTUE

CLAUSEWITZ ON WAR

I turn next to a contemporary of Hegel's who is more securely on the front line whenever the subject is war—for history construed as a march still struts in time to his brisk beat.

Karl von Clausewitz and George Wilhelm Friedrich Hegel died the same year, 1831. Both were pre-eminently men of their time, caught up in the statist and nationalist enthusiasms building in the post–French Revolutionary period. As heirs of the Enlightenment, each also participated in the heady promises of reason triumphant, of the human mind brought to bear on men and events freed from the burdens of ancient superstition. In death, however, Hegel and Clausewitz have been separated: the one to enter the "canon of political thought"; the other to become the *éminence grise* of military history. This separation makes little sense. Clausewitz and Hegel are both avatars of nation building, endorsing reforms they find necessary to carry it out successfully. Political reform, for French Revolutionaries as well as for German champions of powerful, unified nation-states, goes hand in hand with national mobilization; indeed, mobilizing populations for various purposes *is* the form reform takes.

Clausewitz, for example, supported the abolition of serfdom and the mobilization of a large, national army as a task of the central state rather than a traditional duty of the Junkers, the Prussian land-owning aristocracy. The result was a requirement that all citizens (the clergy and educated, middle-class men got exemptions) serve three years in the army and additional years in a reserve force. In the short run, Prussia and its allies defeated Napoleon in 1815 at Waterloo. Over the long haul, what emerged was a powerful warfare-welfare state in the heart of Europe, Bismarckian Germany, which allotted an enormous share of public expenditure to maintain and supply its highly disciplined army machine.

What Clausewitz added to traditional notions of strategy, and the central reason he belongs *inside* rather than exterior to a horizon fused by concern with women and war, is his insistence that *social* forces and attitudes must be tended to: Are people ready for the sacrifices that war and preparation for war require? Is there a latent capacity for armed civic virtue in *this* people at *this* time that can be tapped for *this*

76

purpose? Clausewitz knew that the era of gentlemanly wars involving what Michael Howard dubs "dispassionate professionals" was a thing of the past; henceforth, popular involvement in war, and the expanded size of the armed forces "which nineteenth-century technology was making possible and therefore necessary," meant that "public opinion became an essential element in the conduct of war."[72] Clausewitz warned that war is chameleonlike, changing its nature from one concrete instance to the next. And that nature cannot be disarticulated from the matrix—the maternalism of this word was chosen advisedly—out of which preparedness for wartime sacrifice will either be forthcoming or negated.

If Hegel is the theorist of the state triumphant, Clausewitz is the architectonic champion of, in his words, the theory of *war itself*. War has an "ideal form" and, to be understood properly, must be grasped as "pure concept."* That is, war has certain timeless elements—violence, political impact, and the vagaries of the play of human intelligence, will, and emotions. Political ends must be dominant over military means—*the* Clausewitzian dictum. But Clausewitz's analysis—even as it requires a boundary between political/military such that one is instrumental, the other teleological—transgresses that boundary repeatedly. In the discourse of Clausewitz, we enter the world of war as politics, politics as war that helped to feed the most bellicist of all centuries, the nineteenth.

It would, however, be a mistake to so lodge Clausewitz in his time that we miss the ways in which his views have lofted upward and been carried forward into our own era. His insight that combat is as much a psychological as a political phenomenon; that war fighting "itself will stir up hostile feelings" even if there have been few hostile feelings at the outbreak of hostilities; his insistence that many emotions get "linked with fighting . . . ambition, love of power, enthusiasms of all kinds"; and his recognition that "courage" must outweigh what one is tempted to call the "normal" reaction to danger—namely, flight—all remain salient to discussions of men *and* battle, men *in* battle, and the role of civilian sentiment in war fighting.[74]†

---

* Bernard Brodie is helpful in grasping Clausewitz's war as an idealist construction that never cuts itself off from a pragmatic "fiber."[73]

† Clausewitz also conjures up the "animal" to explain the "human," when it comes to desires for "revenge and retaliation," in a sentence that, by parenthesizing the animal, takes humans a

For example, taking Clausewitz seriously, together with the collapse of the Second International in 1914, helped the Soviet leadership recognize the strength of devotion to country in all the belligerent nations—hence, the utter impoverishment of the notion that the working class thought of itself as a collectivity *sans patrie* or *Vaterlandlöse* ("without a country"). When Stalin called his nation to arms, the battle cry was not drawn from the discourse of Marxist-Leninism, and proffered in its characteristically stilted, dogmatic style, but a *cri de coeur* from the collective soul of the children of Holy Mother Russia. Holy Mother's sons were prepared to sacrifice "for her." Stalin, in this configuration, got to be an honorary papa or a good uncle protecting the great mother, the Russian soil itself abidingly feminized in the eyes, ears, and heartbeats of her people. (Another interesting note: Clausewitz is acknowledged as "the most important influence on Soviet military doctrine to this day."[76] I consider Lenin's celebration of Clausewitz in the next section of this chapter, in a discussion of intellectuals and mobilized language.)

A WOMAN DEMURS: CLAUSEWITZ'S WIDOW

Which leads, finally, to a Clausewitzian subtext or, better, a prolepsis. The preface to Clausewitz's masterwork was written by his widow, Marie von Clausewitz, her husband not seeing the appearance of his great work during his lifetime. When Marie speaks, it is in the voice of the woman decorously in the background, especially, as she puts it, to readers who "will be rightly surprised . . . for such a work as this." Honoring her husband, she rhetorically diminishes herself and her contribution and, to reassure the reader of her proper placement with reference to *On War* and its author, takes steps to "remove any impression of presumptuousness in the minds of those who do not know me."[77]

She tells the story of a bereaved widow determined that the work of her "inexpressively beloved husband," a work that occupied him "almost completely for the last twelve years of his life," should be

---

bit off the hook: "That is only human (or animal, if you like), but it is a fact."[75] We are invited to breathe a sigh of relief as the interpolation, offhandedly, of the animal allows us to think of men turned bestial, hence to scapegoat animals and to retain the "human" scraped just a bit clean.

published. Indeed, Karl had often told her "half jokingly," "*You* shall publish it," as he was reticent about publishing the work during his lifetime. Disarming the reader further, Marie hastens to prevent mis-interpretation of "the emotion that caused me to overcome the timidity which makes it so difficult for a woman to appear before the reading public even in the most subordinate manner."[78]

She constructs herself as a "sympathetic companion," for she could never regard herself "as the true editor of a work that is far beyond my intellectual horizon." She and Karl have had a happy marriage and "shared *everything*": hence, she is "thoroughly familiar" with his mon-umental discursive task—as a demure witness to events. Marie then gives us the sources of Karl's knowledge and erudition—those men he studied with and respected, a brief rundown of his career, his deter-mination to write even when his energies were taken up in or on the battlefield itself. His sudden and tragic death, upon his return to Breslau in November 1831, left her with sealed packages, which she has pub-lished "exactly as they were found, without one word being added or deleted." She goes on, "Nevertheless, their publication called for a good deal of work, arranging of material, and consultation, and I am profoundly grateful to several loyal friends for their assistance in these tasks." Marie's brother has been also a big help. The departed is once more noted as "beloved," and Marie reaffirms that she has been "pro-foundly happy at the side of *such* a man," the italicized "such" signifying that Karl was no ordinary bloke. Marie ends by thanking "a noble prince and princess" for her new job (she had been appointed governess to Prince Friedrich Wilhelm, later Emperor Frederick III).[79]

Marie's few pages are a deft rhetorical performance designed to put *her* in her place or, rather, to reaffirm the place she was in with respect to her husband, the master of war and its discourse. But do we quite believe it? If she and Karl shared *"everything"*—another italicized word—did they not share discussions about the work, its aims, its constructions, its frustrations, its intended audience?

In raising this question, my aim is not to suggest that Marie got gypped out of credit that was rightfully hers but to do something rather different: to ponder why she went through such effort to insist that she had nothing to do, really, with her husband's work. She un-derstood that for a text on and about war, the proper place for a woman—discursively and actually—is in the background, as helpmeet and mirror, honoring the warrior-writer as he sets about his vital tasks. But the fact that she paid such considered obeisance suggests that perhaps the mirror was cracked here and there, that, by 1831, male and female were out of focus with reference to one another's "correct placement" in war—and other things.*

The Revolutionary Alternative: Marx and Engels

If the discourse of armed civic virtue, nation-state style, attains its apogee in Hegel's political thought and in the bellicist excesses of the First World War era, a potent alternative offers its own version of struggle, its own images of historic victories and defeats. One characteristic of the dominant forms of nineteenth-century socialism and communism is the promise of a future "peaceable kingdom" as a real state of affairs, a culminating and continuing moment assured by the working out of the laws of history through an impersonal dialectic. History might require a helping hand here and there by a universal, oppressed class aroused to awareness of its exploitation. But until history has worked its way, violent struggle, including war, is inevitable and necessary: such is the teaching of Marx and Engels.† Communism would bring—indeed would *be*—the form that the earthly peace of humankind in history takes. But, in the interim, there are enemies to expose, foes to fight, whole peoples to triumph or to vanish. War is

* That this issue is no dead letter, or petrified preface, was made clearer than one might have wished during the 1984 presidential campaign when Geraldine Ferraro denied that as Mrs. John Zaccaro she had any knowledge of her husband's business and financial dealings—even though she was vice president of his company. The savvy political in-fighter could combat a campaign liability only by reassuring us, as Ferraro, that Zaccaro was blissfully not in the know. This worked—sort of—because the boundaries between the professional woman "standing on her own" and the wife being supportive and supported are frequently transgressed, and those transgressions hotly denied, in the interest of feminist constructions of a certain sort, in contemporary American society. Marie would have understood—sort of.

† I am not particularly interested in whether and to what extent Engels truly represented positions Marx had only hinted at; rather, I aim to take the measure of their discourse as it affected subsequent representations of violence, particularly theories of social struggle and change.

required to eliminate war. War is not a Wilsonian wedge to make the world "safe for democracy"; rather, it is an impersonal instrument that makes the world ready for revolution.

In her essay, *On Violence*, Hannah Arendt asks what historic transformations and discursive practices made possible an unsettling consensus "among political theorists from Left to Right . . . that violence is nothing more than the most flagrant manifestation of power?" Her answer is multiple. Although she indicts features of several traditions (among others, notions of absolute sovereignty; command-obedience conceptions of law; the intrusion of biologism into political thought), she is especially biting in her judgment of Marxism on this score. Its "great trust in the dialectical 'power of negation,' " Arendt argues, either soothes its adherents, or mobilizes them, into believing that evil is "but a temporary manifestation of a still-hidden good."[80]

Instances of this "great trust" leap out from the constructions of Marx, Engels, Lenin, and Mao. Both Marx and Engels accepted war as a fact of life. Both—especially Engels, an avid student of military history and the "science of war"—admired Clausewitz. Both supported various wars in their lifetimes: for example, the Franco-Prussian War; the French conquest of Algeria ("a fortunate fact for the progress of civilization," wrote Engels); the American war against the "lazy Mexicans"; and any and all Germanic attacks against Slavic peoples on the grounds that Slavs "lacked the historic requisites for independence." Engels put quite pithily the dialectical mission of peoples not destined for historic triumph: their task is "to perish." He and Marx reserved their applause for world-historic peoples.[81] The dream of peace is postponed, and war as a weapon of historic transformation is either accepted as inevitable or endorsed as desirable. Although neither Marx nor Engels espoused a version of *civic* virtue, armed or disarmed, they did embrace visions of revolutionary virtue, the zeal to act collectively and, if need be, ferociously to make class warfare.

Marx believed that a general European war would be required before the full revolutionary effectiveness of the European working classes could manifest itself.* Although scholars and partisans differ on the

* Although Engels took this notion further than Marx, Marx, too, was prepared to extol collective violence as one of the means of history's advance toward stateless universal peace. It

"violence quotient" in Marx, few are prepared to portray Lenin as a spokesman for the peaceable kingdom. For Lenin, the First World War exemplified the cunning of history. A Europeanwide conflagration presaged new and revolutionary understandings that would pave the way for class war. He attacked pacifists as pathetic, *petit-bourgeois* "bacilli." But he also warred on other celebrants of violence. The anarchists and the Blanquists, a French Revolutionary faction, earned his opprobrium for their useless, sporadic "hooliganism." Instead, he promoted and preached collective action on the part of the working class up to and including prolonged civil war embroiling whole countries.[82]

More important to a study of the continued resonance of warrior and warlike myths and symbols than any catalogue of Lenin's embrace of violent historic struggles, however, is his creation of revolutionary discourse as a form of mobilized rhetoric. Via Marx, Lenin picks up on Hegelian tropes and metaphors, decocts them further, honing complex theoretical constructions into blunt instruments, weapons designed to defeat enemies. (And all *opponents*, whether within or without the socialist movement, are *enemies* in Lenin's scheme of things.) The complexities of the Hegelian world, with its moments of ethical particularity (the family) and individuality (civil society) are expunged. Nor are any of Rousseau's ambivalences concerning his own constructions to be found in Lenin's writings with their narrowly circumscribed, purely instrumental purposes.* Lenin's unshakability is a *structured* understanding, a mode of identity, a way of thinking, a blueprint for action made possible by the mobilized discourse he first gives its marching orders and then sends into the world.

Lenin absorbs Marx's most violent locutions: "breaks the modern state power," "parasitic excrescence," "amputation," "destruction." He "takes" the "advice" of Marx and Engels, having first located them

---

should be noted that Marx's revolutionary rivals, Mikhail Bakunin and Ferdinand Lassalle, were not finicky on the subject of necessary violence either, although they *located* it differently.

* The expressivism of Rousseau's *Reveries of a Solitary Walker* (or *Confessions*) are not only unthinkable from the mind and pen of Lenin; they provide irrefutable evidence of the lapidary sickness of the bourgeois soul of even so great a thinker as Jean-Jacques. Lenin was, in Freud's words, one of those "men of action, unshakable in their convictions, inaccessible to doubt, without feeling for the sufferings of others if they stand in the way of their intentions." Freud added, "We have to thank men of this kind for the fact that the tremendous experiment of producing a new order . . . is now actually being carried out in Russia"[83]—but was dubious about the outcome.

as dual authoritative advisers, to "purge" Marxism of "distortions" in order to "direct more correctly the struggle of the working class for its liberation." "Truth" and "power" fuse: there is a single truth, the truth of the dialectic, scientific and exact. Any alternative to this truth/ power nexus must soften the notion of truth constitutive of "scientific socialism": for this truth legitimates absolute and final judgments on persons and events as the prerogative of those individuals who have "correct" understanding. All points of view are judged as either "fundamentally incorrect" or correct. Those who "distort" Marx are thrashed: the potential list—not unlike Robespierre's host of hypocrites—is endless, and they share the same fate: they must be sought out and destroyed. Lenin *precisely*—his claim—locates the inevitability of violent revolution in the writings of Marx and Engels and then reaffirms this inevitability "correctly" as no mere "declamation, or a polemical sally"; rather, the "necessity" of fostering a consciousness aimed at violent revolution "lies at the root of the *whole* of Marx's and Engel's teaching." It follows that "the replacement of the bourgeois by the proletarian state is impossible without a violent revolution."[84]

Just as, for Hegel, war is required to bring into being and to test the state's identity in a world of alienated (potential) rivals, so for Marx and Engels, violent revolution is required to bring into being the historic identity of *the* revolutionary class whose mission is given in advance as the total destruction of the bourgeois state. This, at least, is Lenin's unambiguous reading and re-presentation. Lenin translates Clausewitz's "war as the continuation of politics by other means" into a formula that justifies the elimination of politics through violence, a means Lenin treats as ordinary rather than extraordinary. Because, for him, the "normal" condition of bourgeois society is an undeclared war with winners (the bourgeoisie) and losers (the proletariat), a move to outright warfare only makes the implicit explicit. Any moralizing about this means to revolutionary ends is, on Lenin's view, the petty hypocrisy of sentimentalists drawn from the dominant but doomed class.

When the chips are down, hard-line realism (taken up in the next section) and Leninism can be found holding hands aimed at creating a defensive perimeter that excludes any and all "moralizers." (One is tempted to add "especially women" because they are linked to activities that jar with a world of hostile monads or warring class enemies.) Just as Hobbes bequeathed a vision of a "war of all against all" that dominates the discourse of realism in international relations, so Lenin packaged and mailed to future generations a picture of historic reality and promise that calls for the creation of vanguards and hoists a revolutionary banner that can be kept upright and waving in the breeze only through frequent transfusions of bloodthirsty rhetoric. Thus, the Spartacists—a small, militant band—of the 1970s (or any time since): "the immediate task of our party is . . . to call for the formation of a revolutionary organization . . . an organization ready at any time to build up and consolidate the fighting forces suitable for the decisive struggle." One day, a "Leninist combat party" will "lead the working class in smashing the capitalist order of the bourgeois state."[85] The Weather*persons* of the 1970s saw themselves in this same light, the tag itself changing from *Weathermen* to *persons* in line with women militants' demand for equal billing.

The Leninist revolutionary is a hardcore believer, a don't-look-back actor, tough, thick-skinned, prepared to sacrifice and to struggle over the long haul. Although this representation owes much to received notions of the warrior, the militant is not presented as the exclusive property of male actors. Marx never concerned himself in any sustained manner with the problem of the ways in which the diverse social locations of men and women might lead to differences in self-identity, even among male and female members of the same class. Engels tended to the problem by arguing from analogy that the subjection of women in reproduction is a form of class oppression. He drew women into the picture but in a terribly abstract way, preaching an end to women's oppression as part and parcel of an overall solution to class antagonisms.[86] Nowhere did Engels address whether women are to be perpetrators of violent class struggle, and he certainly didn't think in terms of any identifiably female contribution to socialist thinking and practice.

Lenin was surely seeing males when he imagined "revolutionaries"— but he does not preclude female involvement. He expended more energy thumping female interest in such deviations as psychoanalysis than in detailing any specific, or general, female role in class mobilization and struggle. In the tracts of Mao, however, and given the Chinese leader's concept of "protracted" guerrilla warfare, women come into their own as revolutionary subjects. Mao's is a recipe for total political mobilization, "extensive and thoroughgoing." Although males may be the primary combatants, the attacking prongs, females are absolutely essential to the mobilization effort, to the fight for "perpetual peace."* "War is the continuation of politics," quoted Mao as Clausewitz was reborn again, this time around to shore up Mao's proclamation of an *identity* rather than a relation between the two: "war is politics and war itself is a political action. . . . War cannot for a single moment be separated from politics."[87]† And, in rhetoric mimed by contemporary revolutionaries, Mao asserted:

> Political power grows out of the barrel of a gun. . . . All things grow out of the barrel of a gun. . . . Some people ridicule us as advocates of the "omnipotence of war." Yes, we are advocates of the omnipotence of revolutionary war; that is good, not bad, it is Marxist. . . . We are advocates of the abolition of war, we do not want war; but war can only be abolished through war, and in order to get rid of the gun it is necessary to take up the gun.[88]‡

Jean-Paul Sartre's rationalization of turns to violence and moral absolutism in movements for social change in his introduction to Frantz Fanon's *The Wretched of the Earth*,[90] and Che Guevara's call for revolutionaries to be cold-blooded killing machines, additionally foreclose discursive space within which men and women might take on identities other than, or in addition to, those of mobilized combatants or the hapless others the mobilized must destroy.

Just as a great European war once promised cleansing and redemption to intellectuals of the left and the right on the eve of the First World War, so involvement in revolutionary war evokes the riveting possibility of salutary bloodletting to those who play or work at revolution. Although Lenin and Mao would have anathematized the prewar Futurists as bourgeois decadents playing with fire, they shared much. Lenin and Mao scientized and strategized their rhetoric of vi-

* Had Mao read Kant? Probably not, but he imbibed the locution (see chapter 7 for a discussion of Kant's essay *On Perpetual Peace*).

† That war cannot be "separated from politics," Clausewitz would agree; that war *is* politics, an identity, not a relation, he would reject.

‡ Arendt attacks Mao's conviction as "entirely non-Marxian" because it places violence in a pre-eminent rather than a secondary role.[89]

olent struggle; the Futurists aestheticized theirs. But in the First Futurist Manifesto of 1909, as in the contempt of Lenin for the bourgeois world, one uncovers a common vein—a conviction that destruction must be ruthless and total if construction is to reach its potential for transcendence. The Futurists proclaimed that there is "no more beauty" save in struggle; that poetry "must be conceived of as a violent attack on unknown forces"; that war must be glorified as "the world's only hygiene"; that "moralism, feminism, every opportunistic or utilitarian cowardice" must be destroyed.[91]* The mobilized rhetoric of revolutionists, whether communists or aesthetes, is one of the historic formations—distorted and truncated but all the more powerful for that—radiating out from the darker features of the discourse of armed civic virtue. A second formation—less adrenalized, more conducive to writing monographs than to throwing up the barricades—is nonetheless vexing and dominant in its own right; and it is to this academic offspring that I now turn.

The "Science" of War and Politics: International Relations Becomes an Academic Discipline

Balance-of-power theory assumes that states desire survival and security in an anarchic arena in which self-help, high risk, with the modes and means of competition determined by the most powerful, are the hallmarks of the field. There need be nothing more or less complicated than a desire for security in a structure lacking an overriding authority in order for states to seek the balance of power, or for that balance of power to exist.
—From my Ph.D. qualifying exam, Brandeis University, 1971; subfield, international relations

This is the way I learned it.

A student of international relations, or IR, I absorbed the dominant tradition, cheered on (my caveats, unorganized, being located at the

---

* Feminism is, on the view of the Futurists, but one form of bourgeois decadence, stifling and moralizing.

level of private disgruntlements and mumbled pejoratives) the discourse that had won the war. It was called realism, and it had roots.* These roots were, according to my class notes: Thucydides, Machiavelli, Hobbes, and Rousseau; the Federalist Papers numbers 4, 5, 6, 8, 11, 15, 16; Friedrich Meinecke's *Machiavellianism* (1924); Reinhold Niebuhr's *Moral Man and Immoral Society* (1932); Hans Morgenthau's *Scientific Man versus Power Politics* (1946); George Kennan's ("Mr. X") "The Sources of Soviet Conduct" (1947) and *American Diplomacy: 1900–1950* (1952); culminating in the emergence of a demarcated sub-discipline of political science, a discourse of, by, and for professionals: IR.

The alternative to this discourse of "realism," and its professionalization, was (what else?) "idealism," a potpourri of the writings of religious pacifists, of some if not all just-war thinkers (Augustine slipped into the realist camp on many readings), of liberals (Kant), of natural-law types (Grotius), and a smattering of world-federationalists, one-worlders, peace-through-understanding naïfs, and contemporary behavior-modification enthusiasts out to deprogram the human race away from aggression—most well meaning, very few compelling, but all falling hopelessly wide of the mark where the matter of "why war?" is concerned.

I was already primed for the alternatives of realism versus idealism, having learned in undergraduate school that there are *status quo* states (us) and *revisionist* states (them: specifically, the U.S.S.R.); that states are *sovereign* and have *interests, goals, policies,* pre-eminent among them being *security, power, prestige.* I also learned that the causes of war are inherent dangers in "any international system in which the nation-state remains the arbiter of its own interests and judge of the means by which its security is best assured."†

Realist thinkers and partisans exude the confidence of those whose point of view long ago won the war. Realism's hegemony means that alternatives are evaluated *from the standpoint of realism*—hence, the

* The reader may want to hark back to the seepage of the IR question into chapter 1.

† Although I intend to have a bit of fun with this topic, I am in no way mocking two of the best teachers I ever had: J. Leo Cefkin, Colorado State University; and Kenneth Waltz, Brandeis University. Indeed, I wouldn't know enough to begin to question realism as academic discourse if I had not first learned the discourse—and studied alternatives to it—under their incisive tutelage.

bin labeled "idealism" which, for the realist, is more or less synonymous with dangerous if well-intentioned innocence concerning the world's ways. Realism also promises to spring politics free from the constraints of moral judgment and limitation. In a system of international anarchy (the realist *fons et origo*), wars will occur because there is nothing to prevent them. Force is the course of last resort, and no state can reasonably or responsibly entertain the hope that through the actions it takes or refrains from taking it can transform the wider context.

Modern academics call this "systems dominance." For Hobbes, it was a state of nature for which no cure existed, a never-ending war, as I have said, "of all against all." There is, to be sure, a "logical" solution to this unhappy state of affairs: namely, the creation of an awesome, all-powerful unitary international order, the world analogue of Hobbes's *Leviathan* (1651). But what is logically unassailable is practically unattainable given the world we live in and cannot escape (here comes the refrain): a world of sovereign and suspicious states.[92]*

Historic realism, as molded into a tradition that required and sanctioned the citing of ancient authorities (Thucydides) as clear-cut progenitors of later practitioners (professors at Harvard and elsewhere), involved a way of thinking, a set of assumptions about the human condition, and a potent rhetoric. The great strength of thinkers located in the canon as forefathers is their historic perspecuity; their willingness to deal with the problem of "dirty hands"; their boldness in offering an orientation to the question of collective violence; their insistence that the *limits* as well as the uses of force be treated explicitly, preferably in a mood shorn of crusading enthusiasms, universalist aspirations, and triumphalist trumpeting.†

But—something happened when realism got pinioned within the

---

* Hannah Arendt pays tribute to the great power of the realist tradition when, in her essay challenging teleologies of violence as politics and power, she offhandedly reinforces a Hobbesian view of the international arena. She writes: "The chief reason warfare is still with us is ... the simple fact that no substitute for this final arbiter in international affairs has yet appeared on the political scene. Was not Hobbes right when he said: 'Covenants, without the sword, are but words'?"[93] Donald W. Hanson mounts what, to my mind, is a not terribly convincing case that Hobbes has been missituated as a founding father of realism, and calls Hobbes's mode of thought "unrealistic," "apolitical," and "transhistorical."[94] All this is plausible but somewhat beside the point if one is concerned with how Hobbes has been appropriated and read.

† I will return to these strengths in the classical realist tradition in chapter 7. The suppression of female-linked imagery in historic realism means that the discourse is restrictive in the ways it constitutes its symbolic and narrative possibilities.

academy: it became palpably less realistic, less attuned to the political and historic landscape than in its classical formulations. Encumbered with lifeless jargon, systems and subsystem dominance, spirals of mis-perception, decision-making analysis, bipolar, multipolar, intervening variables, dependence, interdependence, cost-effectiveness, IR special-ists in the post–Second World War era began to speak exclusively to, or "at," one another or to their counterparts in government service. Unabashedly male-dominated (my graduate school bibliography, over 250 titles, includes only 4 entries by women—hopeless idealists all, who sneaked onto my list as examples of what to be avoided), oriented to state sovereignty, presuming unitary notions of power and national interest, IR specialists got caught up in a wider quest for scientific understanding that came out in such forms as game theory and other abstract models that would, so the story went, "work" if one could just get the parameters right.

I remember plowing through many abstruse exercises. In one whose significance eluded me altogether, the outbreak of the First World War, having been reduced to a finite number of variables, got trans-formed into a model for computer simulation. Would the "outcome" be the "same"? Who cares? I thought—the war's nine million soldiers will stay buried. But evidently I was supposed to marvel at the con-clusion that, if the statesmen at the time had the knowledge available to them of what the outcome was going to be, they might have acted differently!*

Characteristic of modern professional discourse in its most recent incarnations, then, is a proclamation of scientific knowledge; a pre-sumption that politics can be reduced to questions of security, conflict management, and damage control; a patina of "aseptic, ahistorical and anodyne terminology" ("window of vulnerability," "collateral dam-age," "crisis management," "escalation dominance"); and a pronounced insouciance concerning the will to power, including the promise of control over events, embedded in the concepts and tropes that comprise the discourse in the first place.[96]

Although particular forms of this quest for scientific certainty come

---

* I'm trivializing, but not by much. In my camp are heirs of classical realist discourse who found—and find—the rush to scientize pretty foolish, much to their credit. Those who do not identify themselves as realists—for example, Stanley Hoffman—also criticized the developments I am treating as risible even as they were coming into focus.[95]

and go, the dangers inherent in professionalized war discourse remain. I have no better word for what I have in mind here than "dissociation." For although the IR specialist as a constructor of abstract scenarios, cloaked in the legitimating mantle of "scientific study," presents himself as one who describes the world as it is, he is living out a perilous fantasy: the delusion that we have control over events when, in fact, we do not. Arendt argued, in 1969, that "scientifically minded brain trusters" bustling about in think tanks, universities, and government bureaucracies should be criticized harshly not because they were, as some liked to boast, "thinking the unthinkable" but, rather, because "they do not *think* at all."[97] The scientific pretensions of "rationalist realism" eclipse the strengths of the classical tradition, including awareness of the intractability of events and a recognition that relations between and among states are necessarily alienated, more or less estranged.*

As a modern classicist realist, Michael Howard states:

When I read the flood of scenarios in strategic journals about first-strike capabilities, counterforce or countervailing strategies, flexible response, escalation dominance, and the rest of the postulate of nuclear theology, I ask myself in bewilderment: this war they are describing, *what is it about?* The defence of Western Europe? Access to the Gulf? The protection of Japan? If so, why is this goal not mentioned, and why is the strategy not related to the progress of the conflict in these regions? But if it is not related to this kind of specific object, what are we talking about? Has not the bulk of American thinking been exactly what Clausewitz described—something that, because it is divorced from any political context, is pointless and devoid of sense?[99]

That women have been pretty much excluded from this scientific enterprise is not its most obvious flaw, and it is one that can be remedied. There are women, in increasing numbers, prepared to take their place among the ranks of the purveyors of the hegemonic discourse—whatever it may be at any given moment in the academy and its journals of, by, and for the collective "we." Women can be included, can engage in the activity, even if *representations* of women and the sphere with

---

* James Der Derian offers a bold reinterpretation of the discourse of international relations and diplomacy which locates "alienation" as a central category.[98]

which they have been historically linked remains an absence that helps to make possible the much cherished "parsimony" of the preferred model, or framework, or simulation, or analysis in the first place. For example: if one can concentrate exclusively on states and their "behavior," questions of human agency and identity fall to the wayside. No children are ever born, and nobody ever dies, in this constructed world. There are states, and they are what is.

Professionalized IR discourse, whether as abstract strategic doctrine advertising itself as realism brought up to date, or as most alternatives that would take us "beyond realism," is one of the most dubious of many dubious sciences that present truth claims that mask the power plays embedded in the discourse and in the practices it legitimates. That some individuals escape the snare of these modes of professionalization and emerge from their training still able to conjure with the complexities of history, the vagaries of events, and the unpredictability of human passions is a testament to a wider human attunement to *common* sense, shared if unstated recognitions, diffuse if unsystematized understandings. The man, and the woman, "in the street" often knows how fragile it all is, how vulnerable we all are. Indeed, one of the virtues of what Sara Ruddick, a feminist philosopher, calls "maternal thinking" is precisely a recognition—frequently rueful, even grudging—that the mother neither has nor should aspire to total control over the life and development of her child.

Men and women of common sense may hope for profound transformations, may imagine peaceable worlds. Most importantly, however, they mistrust those who, on the one hand, conjure with acceptable death ratios in their scenarios of nuclear war fighting or, on the other, pretend that such disasters are not remotely in the cards. We who traffic in the written word do well to join these prototypical citizens in their skepticism and to take seriously their hopes. The hopes I here evoke have been embodied within a hybrid genealogy whose historic markers include pacifism, just-war doctrine, and several of the many strands of liberalism. But before I turn to these alternative narratives, historic stories of female sacrifice and honor and of twentieth-century mobilization for war, made possible in part through reinscription of the tradition of armed civic virtue, beckon for attention.

3

Exemplary Tales of
Civic Virtue

By oppression's woes and pain!
By your sons in servile chains!
We will drain our dearest veins,
But they shall be free!
Lay the proud usurpers low!
Tyrants fall in every foe!
Liberty's in every blow!
Let us do or die.

—ROBERT BURNS
"Scots Wha Hae" (last stanza)

MICHEL FOUCAULT links "the fact of killing and the fact of writing, the deeds done and the things narrated," finding each—deed doing and text writing—forms of authorship, of assertion of the right to kill and to narrate.[1] While this puts the matter rather starkly, various narrative forms and theoretical constructs *do* call upon and help constitute diverse understandings of the self, male and female, in relation to the *civitas*, whether as local *patria* or Hegelian *Kriegstaat*.

The narrative of patriotic armed civic virtue, as I observed in the preceding chapter, constructs the woman either as Machiavellian mirror to male war making or as the exemplary and stalwart mother of Rousseau's Spartan ideal who loses five sons but gives thanks to the gods that Sparta won the battle. The Rousseauian vision draws woman more directly into the picture. She is both necessary and integral to the life of the virtuous pre-state body politic. Within the Machiavellian construct, women's dramatic roles are less visible, more privatized. Rousseau, however, represents marriage and family as institutions with an explicit civic purpose. And women have seen themselves through that lens; have told the story of their own lives enframed by this grand narrative.

The Tall Tale of Armed Civic Virtue structures stories of wars and "the woman" by enabling individual women to make sense of what is happening to and through them. The woman of republican militancy is no mere victim of events; rather, she is empowered in and through the discourse of armed civic virtue to become an *author* of deeds— deeds of sacrifice, of nobility in and through suffering, of courage in the face of adversity, of firmness in *her*, and not just her polity's, "right." Just as the soldier is prepared to derealize himself as a civilized being ("Thou shalt not kill") to preserve the civic mother that gave birth to his civility in the first instance, the mother/wife of the soldier is prepared to sever herself from the most potent imperative under which she ordinarily labors: "Thou shalt protect the bodies of thy children."

To preserve the larger civic body, which must be "as one," particular bodies must be sacrificed. The soldier is called to forfeit his own. The mother is called to forfeit those of her sons; the wife, that of her husband. To the extent that she has become a civic person, to the extent that the other imperatives (most importantly, Christian strictures against killing and violence) that make her who she is can be at least temporarily jettisoned, overridden, or themselves mobilized for combat, she is available as a civic republican mother whose call is not to arms but to dis-arm, to break the protective enfolding of her children, to cherish "public freedom" above "private devotion."

She must be tough. Examples of such toughness rise to the surface in various epochs. In this chapter, I shall focus, first, on a historic moment that exemplifies women's social location as civic republican mothers, the daughters of Sparta, their militancy softened somewhat by moral queasiness concerning violence and by reigning views of motherhood. This moment is the Civil War, North and South.

Women's actions and reactions were central to the war effort on both sides.

The war to test the concept of union having been won (at terrible cost to both sides) by the North, America nevertheless remained a society devoted to local and regional identity. This plethora of particularities was enhanced dramatically by the great waves of immigration of "foreigners" to American shores in the late nineteenth and early twentieth centuries. Many political actors and thinkers fretted obsessively about what immigration meant to America. The solution proved to be the First World War. A *united* states finally emerged. The wrenchings of that epoch, the second historic moment on which I shall concentrate, illustrate in often dreadful detail the political requirements of armed civic virtue. Here, too, women played their part or were compelled, as were men, to join the crusade or be ostracized.

Women and the Civil War

SOUTHERN WOMEN: "MY COUNTRY RIGHT AND WRONGED"

Oh yes, I am a Southern girl, and glory in the name;
And boast it with far greater pride than glittering wealth or fame.
I envy not the Northern girl, her robes of beauty rare,
Tho' diamonds grace her snowy neck, and pearls bedeck her hair.
Hurrah! Hurrah! for the Sunny South so dear!
Three cheers for the home-spun dress that Southern ladies wear.

—Sung to the tune of "The Bonnie Blue Flag"

In one of John Ford's minor classics, *The Horse Soldiers* of 1959, John Wayne as Colonel John Marlow leads his soldiers into the Southern homeland. A hardbitten man-at-arms, he's got a job to do and he'll do it well (though it is clear that he's nursing some private, unexpressed wound). His troop of horse soldiers is accompanied by one Dr. Kendall (William Holden) who embodies, at least initially, the compassionate body healer as compared with Marlow, the no-nonsense trooper who knows no god save military necessity. Seeking to sequester his men, Colonel Marlow homes in on a Southern plantation presided over by a suspiciously sweet hostess ("Welcome to Greenbriar, gentlemen," she croons venomously). In private, the lady of the mansion curses the Yankees to her confidante, her black "servant," in language that expresses her true feelings: they are "nameless, fatherless scum."

When Wayne/Marlow discovers the mistress of Greenbriar trying to spy on conversations between himself and his officers, he has "no choice" but to take her hostage. She must ride along with him and his men or she poses a risk to their mission. Entering a Southern town deserted by its soldiers, the Yankees encounter an angry mob of Southern women who line the streets and throw dirt and rocks, shouting epithets and screaming curses. The town, Newtown Station, becomes the site of a battle that Colonel Marlow has tried to avoid ("I didn't want this"). Turns out he's a softy after all, who is sickened by the bloody business in which he is engaged. Anguished at the destruction, he gets drunk, proclaiming, "I build railroads." In the meantime, the Union doctor and the Southern lady nurse the wounded from both sides: the brotherhood of death overcoming sectional hatreds. One young trooper calls for Colonel Marlow and dies in his arms: "Just hold on to me, sir, and write my Ma."

Marching out of the town, seeking to avoid another fight, Marlow faces the horrible prospect of pitting his troops against the only Southern "men" left: the boys of a military academy. One mother begs the unarmed commander (the minister-headmaster of the school) to allow her son to remain home. She grabs her son, but he resists and escapes to do battle. But Marlow avoids the fight; the schoolboys are unharmed; the doctor is disarmed, realizing Marlow is a gentle(man) beneath the crusty soldier's surface; and the Southern woman who, by this time is hopelessly in love with Marlow, weeps as he departs with apologetic words for the "hardship and humiliation" he has caused her.

The film is bursting with the semiotics of culturally familiar war imagery. What jolts the viewer is the scene of all those angry women damning the invader. We (the collective American movie audience) *know* from the greatest of all Civil War films, *Gone with the Wind*, that individual Southern ladies could hate the Yankees when they literally invaded the home: that memorable moment when Scarlett

O'Hara blasts a scummy, obviously debased Yankee cur, saving herself and her Negro "servant." So we're not surprised by the lady/spy/hostage/nurse/can't-help-but-fall-in-love-even-though-he's-one-of-them protagonist of *The Horse Soldiers*. But that mob of militant women, who have poured *out* of their houses and into the streets, is not among our stock set of images. No individual heroine, beautiful but prepared to defend her domicile; rather, a female mob *attacking* with words and stones a troop of horsemen whose expressions convey shock, even horror, at the hatred of these women *toward* them. That's an image we are less familiar with, though it conveys a previously narrated truth: the determined militancy of Confederate women during the Civil War.

Northern women were patriotic, too, and I shall retell a few of their stories later. But Southern women best exemplify the female pole of civic republican virtue for several reasons. First, they were fighting to defend a way of life that incorporated exclusivities—indeed, was erected on that basis. By this I mean that the civic republican tradition has always *limited* citizenship and has defined who is *within* and who is *without* the *polis*, who is necessary to but not, strictly, an integral part of it. The black slaves who made possible the Southern way of life were *not* expected to march to war to defend it—not only because that was the task of its explicit warrior/dominant class but because of fear of slave revolt.*

Second, Southern women defined their *patria* locally—as a way of life they had no particular desire to extend but a keen stake in maintaining. Northerners represented a more universalist, democratic prospect—extending citizenship to previously excluded groups; inscribing a sense of an inclusive union upon all who fell under the aegis of that nationalizing instrument, as the Federals interpreted it, the Constitution; ending sectionalism by eroding the bases of Southern identity which included the fierce pride of an educated landed class and its slaves—though Southern women in their memoirs speak only of "servants" and fondly, though matronizingly, at that.† The South made

---

* These features of the pre-war South suggest a mimetic relationship to narratives of classical antiquity.

† Some Southern women lamented the institution in which they shared. Mary Chesnut, in her war diaries, describes "a sale of Negroes," of "Mulatto women in *silk dresses*—one girl was

war upon the army of the North; the North made war upon the people of the South: that, too, signifies the North as modernizing and offensive, the *nationalistic* as compared with *patriotic* force (a distinction I shall explore in greater length in chapter 7).

Mary Boykin Chesnut, a Southern lady, began her war diaries in February 1861 and crafted them with an explicit narrative purpose, a public aim: staying within the diary form, she nonetheless turned to tropes and locutions, to literary genres of public expression. We learn, from her diaries, of hardships endured: "Fasted yesterday all day. Not one morsel I tasted all day"; and, just one sentence later, of similar resoluteness in the publicly armed, not only privately sacrificial, aspects of the struggle: "A ship has got into Savannah with *arms* in plenty for five brigades. God be thanked." She laments the wounds incurred by "our poor country. Oh my poor country"—but would rather "that we might all be twenty feet under ground before we were subjugated."[3]

Chesnut curses Lincoln, the enemy leader ("an insidious villain"); mourns her husband's going to war ("makes me miserable!"); and can locate her feelings only by citing a long passage from Tennyson's *Idylls of the King*:

> And yet I hate that he should linger here;
> I cannot love my lord & not his name.

The worse fate to befall her would be "that my Lord through me should suffer shame," this final line by Tennyson being followed with her own words: "So feeling, I used not one word to prevent his going—because I knew he ought to be there."[4] Another of the many poems Chesnut quotes to situate her memoirs inside the long literary tradition of men-to-war, women-to-suffer is Richard Lovelace's "To Locasta, On Going to the Wars," whose protagonist, in a precise reversal of Chesnut's Tennyson-constructed female heroine, tells his "sweet" that he must fly "to war and arms," an "inconstancy" that she "too shall adore," for "I could not love thee, dear, so much/Loved I not honour more."

---

on the stand. Nice looking—like my Nancy—she looked as coy & pleased as the bidder. South Carolina slave holder as I am my very soul sickened—it is too dreadful. . . . The Bible authorizes marriage & slavery—poor women! poor slaves!"[2]

At one point, Chesnut reinscribes the words of Hecuba—who mourns her grandson's *miserable* death* by dignifying the *grand* death he might have had fighting for Troy—in railing against a woman who "utter[ed] the basest sentiments I ever heard fall from the lips of woman. I said 'dead upon the field of battle—dead—out of the way of shame & misery' was the best. *'To fill a patriot's grave the noblest fate a man could crave.'* She asked me if I imagined all the men who filled patriots' graves were going to Heaven—& spoke so heartlessly & flippantly" (italics added).[5] The craven female Chesnut assaults might, from another angle of vision, embody the voice of the female skeptic, one suspicious of the notion that God guides the guns of all dead patriots. But within Chesnut's civic republican narrative, rounded out and filled in as it is with literary allusions and historic exemplars that help her make her case and plead her cause, her female protagonist *must* appear insensitive and foul, both weak-willed and cynical.

Chesnut is unusual *only* in her literary construction of herself against the backdrop of events she saw as grand and tragic. Histories incorporating first-person accounts by Southern women detail their heroism, their ingenious capacity for providing substitutes for goods made scarce by the Northern war of attrition, and their determination to sustain and conserve "life at home," enabling "the Confederate forces to keep the field." "We are very weak in resources, but strong in stout hearts, zeal for the cause, and enthusiastic devotion to our beloved South," wrote one woman; "and while men are making a free-will offering of their life's blood on the altar of their country, women must not be idle." By all accounts, the women were not. "We must not admit weakness," noted a Mrs. McGuire in her diary. "Let States, like individuals, be independent—be something or nothing."[6]

One "Mrs. Betsy Sullivan" became "Mother to First Tennessee Regiment." Childless, Mrs. Sullivan not only accompanied her husband to war but also set out to "mother" an entire company. The writer who honored her noted that the men would dare all for her because "her presence represented to them—their wives, their mothers, and their homes. In turn, Mrs. Sullivan held, in her long years of hardship with the army, that no trial was too severe, no sacrifice too great, if

* Too young to fight, he was thrown off the city's walls and killed.

98

made on behalf of her 'boys.' "* One Veteran of the First Tennessee affirmed to the author of *The Women of the South in War Times* that "not one single man in the entire regiment would have hesitated to spill the last drop of blood for 'Mother Sullivan.' "[8]†

Heroines spill out from the pages of hagiographical works: Mrs. William Kirby, captured and imprisoned as a blockade runner; Mother Philips, another childless woman who followed "the fortunes of her husband through weal or woe"; the many women who pioneered that organized succor later institutionalized as the "noble army of the Red Cross," who gave themselves over to sewing uniforms, making bandages, putting up food, running canteens, nursing the wounded, offering their own stores of essential supplies to keep the soldiers marching. The authors of *The Women of the Confederacy* detail instances of female heroism under the rubric "Manifestations of Spartan Motherhood."‡ And manifest they were: women who traveled into enemy territory to find their wounded or dead husbands; the "aged mother" who told her widower son, hesitant about "joining the army in Alabama because of a desire to avoid placing the burden of the care of his children" upon his mother: "Go, Jack, the country must have men, and you must bear your part, and I will take care of the children." Jack, worried about his children in case of his death, remonstrated with his Spartan Mom but she, in turn, declared (the tone of impatience rising in her voice): "Jack, I will do a mother's part by them; but you must not talk that way." Another story is told of the "humble Virginia woman" whose sons and husband were all in the service: "Oh yes, I shall miss my husband mightily; but I ain't never cried about it; I never shed a tear for the old man, nor for the boys neither, and I ain't

* Should the reader find this a curiosa of one particular historic configuration in nineteenth-century America, an example drawn from another culture, another time and place, may help demonstrate the abiding robustness of these companion images of "fighting for" and "sacrificing in behalf of." Left-wing Peronistas in Argentina conjured up a dead Eva Perón as inspiration. She called them to battle: "When we are discouraged, we draw our strength from Eva Perón. ... She exemplified personal force at the service of the revolution: the dedication to the process of change, to the accomplishment of a goal. ... We are fighting because of Evita—and for her."[7]

† The example of Mother Sullivan and other women who, having no children of their own, set about becoming maternal beings in a wider sense, indicates that biological motherhood and social maternalism can, and have been, teased apart—to fight wars and to promote peace (see the discussion of Jane Addams, another maternal figure, in chapter 7).

‡ Plutarch's "sayings of Spartan women" echo and re-echo in the voices of Confederate women.

agwine to. Them Yankees must not come a-nigh to Richmond.""⁹

Compilations of muster rolls honor two North Carolina mothers who gave eleven sons each to the Confederacy; four who gave nine sons; two, eight; nine, seven; eighteen, six; nineteen, five; thirteen, four; twenty-two, three. The spirit animating such giving up and over of the bodies of one's children was expressed to a newspaper reporter interviewing one of these Spartan exemplars: "I have three sons and my husband in the army. . . . They are all I have, but if I had more, I would freely give them to my country." A Tennessee mother, having lost three sons, gets ready to send "Harry too." The authors, their spines stiffened by the female toughness they, in turn, reify, intone: "Exalted suffering and sacrifice demanded an exalted vengeance from these Spartan mothers. The blood of the slain sons called for additional sons to battle for the vindication of those who had fallen." These are female figurations Rousseau would recognize, and honor. Struggling for the independence of their way of life, Southern women cursed the foe, agitated the homefront, "rushed out of their homes" to champion the Confederate cause by stimulating enlistments (one historian notes that "the cowards were between two fires, . . . the Federals at the front and . . . the women in the rear")¹⁰; creating relief and soldier's aid societies; providing individual and collective examples of martial enthusiasm and religious faith in the Southern cause; and receiving the "enemy" with hatred and invective.

The authors of *Women of the Confederacy* up the rhetorical ante in a struggle to find words adequate to honor this republican heroism: "Had the women of the Confederacy lived in a more heroic age, in which their sex put on armor and went forth to battle with dignity, it is possible that they would have produced real Amazons and Joans of Arc. Such conditions, however, did not exist in the middle of the nineteenth century, and the Southern women had to express the most of their heroism in less spectacular forms."¹¹*

That the stories of armed virtue in the words and deeds of its female prototypes is no simple tale of mythological construction by interested

---

* These habits of civil-war time seem to have stuck: in 1917, when the call to arms in behalf of the United States of America went forth, Southern women under the auspices of the United Daughters of the Confederacy, some sixty thousand strong, put themselves at the service of this national cause by engaging, in disproportionate numbers, in war relief of every sort.

parties is evident if one considers Northern war strategy as played out in General William Sherman's March to the Sea. Immersed in the "war is hell" ethic, Sherman, according to James Reston, Jr., disdained standards of "proportional response and discriminating protection of civilians":

> [Sherman] expressly set out to make Georgia howl. But neither states nor soldiers howl; civilians do, particularly women. It took someone who knew the South and Southern pieties well to understand just how effective making war on women could be. The problem with historical writing about the end of the Civil War is that its language is grandiosely and deliberately imprecise. Sherman would break "the will of the South" to fight, *but his technique was to demoralize the women back home, and let that have its effect on the soldiers at the front.*[12] (Italics added.)

Had Southern women not constituted themselves as Spartan mothers, the Sherman strategy would have been unnecessary. Understanding that the will to resist turned on the support of the civilian (female) population for the war effort, Sherman's strategy of (near) total war obliterated any distinction between the battle front and the home front, a distinction Southern women themselves actively disdained.

A monument to Confederate women, on the grounds of the South Carolina State House plaza, has inscribed upon its pedestal words that a twentieth-century anti-war writer found "unexpectedly moving . . . for all their old-fashioned floridness":

> Their unconquerable spirit strengthened the thin lines of gray. Their tender care was solace to the stricken. Reverence for God and unfaltering faith in a righteous cause inspired heroism that survived immolation of sons and courage that bore agony of suspense and shock of disaster. The tragedy of the Confederacy may be forgotten, but the fruits of noble service of the daughters of the South are our perpetual heritage.[13]

These words are moving but not unexpectedly so. They touch us because they conjure up images of collective *esprit*, tapping a deeply rooted cultural double-mindedness concerning war and its demands. On the one hand, we see the (honored) horror of giving over living sacrifices in the bodies of male children for the survival of the homeland. On the other hand, we see the *opportunity* for women to engage in deeds that partake of received notions of glory, honor, nobility, civic virtue. Whether martial or succoring mother, the words on the base of the statue honoring Southern women are inscribed on our collective civic souls, calling men and women to sacrifice in behalf of a people, a homeland, a way of life.

NORTHERN WOMEN: "OUR TRUTH GOES MARCHING ON"

Northern reverberations during the Civil War echo the Southern construction of patriotic womanhood. Stories of "heroism and self-sacrifice" are detailed, with one compiler of tales, in his introduction, arguing that, in "our Conflict for the Union," women no longer suffered "in silence at home" but were implicated in the very thick of things:

> No town was too remote from the scene of war to have its society of relief. . . . Everywhere there were humble and unknown laborers. But there were others, fine and adventurous spirits, whom the glowing fire of patriotism urged to more noticeable efforts. These are they who followed their husbands and brothers to the field of battle and to rebel prisons; who went down into the very edge of the fight, to rescue the wounded, and cheer and comfort the dying with gentle ministrations; who labored in field and city hospitals, and on the dreadful hospital-boats, where the severely wounded were received; who penetrated the lines of the enemy on dangerous missions; who organized great charities, and pushed on our sanitary enterprises; who were angels of mercy in a thousand terrible situations.[14]

Uttering what we would today construe as a feminist plaint, this Northern hagiographer of women in war insists that the "story of war will never be fully or fairly written if the achievements of women in it are untold." Although women do "not figure in the official reports," their tales should be told, and it is the object of his book to "present narratives." And he does, a gallery of noble Northern women standing metonymically for Northern "womanhood": Mrs. Fanny Ricketts, Mrs. Mary A. Brady, Mrs. Elida Rumsey Fowle, Margaret E. Breckinridge, Miss Major Pauline Cushman, Mrs. Mary Morris Husband, Anna Maria Ross—several dozen heroines "named" and thus officially "deeded," acknowledged as authors of a story.[15]*

* More recently, the novelist Rita Mae Brown has launched a similar appeal for recognition

Exemplary Tales of Civic Virtue

The Spartan imagery is invoked for Northern understanding: indeed, American mothers "with more than Spartan patriotism, sent forth their sons to fall by rebel bullets, or to languish in rebel prisons." Mary Brady's tale is reinscribed as an updated version of Greek heroism at the time of the Persian invasion. Quoting "an old Greek writer" (Heraclitus) to the effect that "War is the father of all things," Frank Moore solidifies the North's relation to the ancients with a tidy segue: "In like manner, we of America, looking at all the latent heroism that was developed during those four years of national agony and national glory, may ... hail our great war as the father of a great national peace."[17] The scarcely muffled crow of victory joins a Hegelian emanation: war is the father of peace; war is the creator of great *nations*.

But it is clear that Northern women were more akin to their Southern sisters than to Northern avatars of a vision of a grand nation. They were sacrificing for their *patria*; they were nursing their sons, husbands, fathers; they were defending, as they saw it, a way of life. Not surprisingly, few in paeans to war sacrifice are found exalting anti-slavery sentiments on a grand rhetoric scale. Instead, their words are cast in the stentorian tones of duty and citizenship, self-sacrificing for a greater cause, agitation not so much against slavery as for "the union" for it was "the union" that constituted their larger sense of homeland.*

Images that conflict and cohabit are traced on the pages of the Northern women's story. The frontispiece of Frank Moore's *Women of the War* (see photo section, #8) features—above the words "Before the Battle"—a beautiful young woman in repose, hands clasped in an intimation of prayer, eyes gazing mistily outward, hopeful and yearning yet prepared for the worst. This woman is the young wife, prepared to forfeit her husband (and perhaps her womanhood) on the altar of patriotic necessity. A second engraving, "The Mother's Sacrifice" (#9), almost unbearably poignant, pictures a mother embracing her very young son, ready to march off as a drummer boy. An older soldier, holding the drum and drumsticks so the youth can be enfolded one

---

of "the sacrifice of our women" in time of war. The mother of Brown's heroine, Lutie Chatfield, becomes a tower of strength nursing the wounded on the home front. The protagonist, Geneva, transforms herself into "Jimmy" and goes off to fight.[16]

* This statement must be qualified: many women—and men—saw themselves as citizens of Maine or Massachusetts first, and only secondarily as members of a more inclusive polity.

last time in his mother's arms, waits patiently for the moment to conclude. A third image vies for attention: "Kady Brownell in Army Costume" (#2). Kady Brownell, the wife of a soldier, followed her husband to battle, experiencing "all the hardships of the camp" and devoting herself "with the delicate tenderness of her sex, to mitigating the horrors of the battlefield." But Kady Brownell looks neither delicate nor tender. Allowing for the standard severity of portraiture in this era, Brownell presents herself as the avenger who can hold her own and whose self-designed uniform (shades of General Douglas MacArthur!) signifies her romantic and heroic sense of self. It turns out Kady was an army brat, accustomed to arms and soldiers "from infancy." The outbreak of war gave scope to her own enterprise, but author Moore must still cast it in the sentimental locutions of the day—she cut a "graceful figure on parade"—though he does admit she became "one of the quickest and most accurate marksmen in the regiment," and her sword was no idle instrument for she "practised daily." Re-enlisting with her husband, she stayed in the thick of things, though she seems not to have done any actual killing. Other women, more traditionally (like Margaret E. Breckinridge), knit socks and wrote doggerel about sock knitting "from daylight till dark," ending, "Ah, my soldier, fight bravely; be patient, be true/For some one is knitting and praying for you."[18]

Womanly duties, clearly, encompassed a wide sphere. Out of the *oikia* ("households") and into the *polis*, though the view of the *polis* was often a domesticated one, women saw themselves, and others saw them, through received cultural lenses. Those who, like Kady Brownell, challenged standard figurations were made to fit and seem not to have gone quite all the way into constituting themselves *wholly* as soldiers, though Moore notes that some were "found in the ranks, . . . their sex remaining unsuspected, and the particular motive in each case often unknown."[19]*

Spartan mothers, Cato's daughters, these women of the warring parts of the republic located themselves within a horizon framed by a tradition of which they may have been only partially aware yet that

* In *The History of Woman Suffrage*, Elizabeth Cady Stanton and Susan B. Anthony also celebrate the occasional woman who passed for a (male) soldier, even as they condemn war (see my discussion of the vagaries of feminist approaches to women and war in chapter 7).

spoke and acted through them nonetheless. Reinforced by Biblical notions of the triumph of right, with succor and forgiveness temporarily overridden by just-and-"holy" war notions (actually a contradiction within the just-war tradition as constructed historically, for that tradition forbids—at least rhetorically—explicitly crusading wars of conquest), women acted and wept, playing absolutely essential parts in the grand and terrible drama.* Attuned to Greek precedence, in which hero graves became public gathering places for the *polis*, and to the medieval cult of the saints with its cherished remnants of the sanctified holy sites for Christian pilgrimage, the Union forces in the American Civil War formed battlefield cemeteries with individualized markers for soldiers' graves.[20] War cemeteries and monuments meant the dead got to be actors in civic life, serving to integrate a sense of "peoplehood."

That these Civil War images continue to reverberate is suggested by the thirty-foot female figure that looms in the memorial to the dead of the Pacific Theater, in the Second World War—the Memorial National Cemetery of the Pacific (popularly called "Punchbowl") in Honolulu (see photo section, #10). The visitor mounts dozens of rows of gradually narrowing steps as she slowly approaches the female figure "who stands on the symbolized prow of a U.S. Navy carrier with a laurel branch in her left hand," as the language of the American Battle Monuments Commission, in 1985, put it. What the visitor's guide fails to note is that this female figure is helmeted, offering at once both peace and war. She is strong, with resolute features, the wreathless right arm straight at her side, slightly extended away from the body, palm facing outward—a sign of openness and welcome. But the right

---

* A Soviet variant on Russian armed civic virtue is featured in Sergei Eisenstein's classic 1938 film *Alexander Nevsky*. In the thirteenth century the Teutonic Knights, portrayed as the sinister embodiment of imperialist evil, invade the Russian homeland. Not content to fight the men, "they torture women whose men war against them." A father, dying, to his daughter cries, "Avenge us." Following a glorious call to arms, a mass spontaneous upsurge in civic virtue follows: "Fame to the living/Glory to the slain." Nevsky leads into battle the patriots, including Vasilisa, a fighting woman. Vasilisa is interesting in that she also plays out the part of the "traditional woman," baiting two possible suitors: "Whichever one of you shows the greater valor, he will I wed." After the battle, women carrying torches look for their dead and wounded ("He who fell for Russia in noble death/His dead eyes will I kiss"). The mother of one fighter, Vasily, frets that he may have come in second in bravery and thereby "brought disgrace on your mother."

leg, visible through draped folds, powerfully muscular and firm, is bent in a gesture that suggests forward motion: she will welcome you or come *at* you: the choice is yours. She is slow to wrath but firm in resolve. Engraved below the figure is, in the words of the pamphlet, "the poignant sympathy expressed by President Lincoln to Mrs. Bixby, mother of five sons who had died in battle; . . . 'THE SOLEMN PRIDE THAT MUST BE YOURS TO HAVE LAID SO COSTLY A SAC-RIFICE UPON THE ALTAR OF FREEDOM.' "

Lincoln's words speak to, and of, the solemn pride of civic republican mothers everywhere. The words touch us in the very different context of the Second World War dead, in part because patriotic compared with nationalistic stirrings seems so much a part of that war which lent itself—too easily—to dichotomies of aggrieved good versus menacing evil, of freedom assaulted versus totalitarianism assaulting. Calling upon Lincoln's Mrs. Bixby, stirring old images and identities, shoring them up for a specific moment, burial of the dead of the Second World War in a manner that makes it possible for us to call upon them at some future moment should it be necessary, makes sense: "our truth goes marching on," in the words of Julia Ward Howe's "Battle Hymn of the Republic." That truth, honed through this century's conflicts, brought America closer to the civic republican dream of one nation indivisible, but at a cost to other understandings of ourselves as a nation and a people.

The First World War: "My Nation-State, of Thee I Shout"

We have seen . . . how a divided and arguing public opinion may be converted overnight into a national near-unanimity, an obedient flood of energy which will carry the young to destruction and overpower any effort to stem it. The unanimity of men at war is like that of a school of fish, which will serve, simultaneously and apparently without leadership, when the shadow of an enemy appears, or like a sky-darkening flight of grasshoppers, which, also are compelled by one impulse, will descend to consume the crops.

—EDMUND WILSON, *Patriotic Gore*

The statist nationalism of the First World War, an approximation in Europe and the United States to earlier dreams of millions of civic souls marching in time to the same rousing tune, is both apogee and nadir of nineteenth-century nationalism. Apogee, because whole peoples were mobilized, because the state did come to life, brimming with a sense of unified self, overflowing its boundaries to defend and guarantee that crystallization; nadir, because of the mounds of bodies sacrificed in a prolonged, dreadful orgy of destruction. "Trench warfare," it was called, and it meant mass, anonymous death. In the first day of the Battle of the Somme, 1 July 1916, 60,000 men were killed of the 110,000 on the British side who got out of the trenches and began to walk forward along a thirteen-mile front. The "final figures," civilians exempted, toted up to 8,538,315 men killed in action or dead of wounds on all the fighting fronts, and 21,219,452 were wounded and maimed.*

The question at hand is the bringing into being of grand civic entities in and through war—preparation for war, fears of war, anticipation of war, and actual war fighting. To create a grand civic entity, local identities must be shattered or muted; individuals must become entangled with the notion of a homeland not as the local community into which one is born but as a vast entity, symbolized by flags, oaths of allegiance, constitutions (in some cases), and wars *against* others.

In the nineteenth century, the concatenation of enhanced industrial and technological power and increased consolidation of the means of violence commingled to a given end: the overweening trappings of national sovereignty, with its concomitant and subsequent moves to standardize and pacify the areas under its purview (a process I have already described a bit of, in the French Republic, in chapter 2). Although the idea of *nationalism* predates the creation of centralized nation-states and can exist in their absence, the fusion of the nationalist idea with the nation-state apparatus is a particular historic configuration, one that remains dominant either as a political aim (for pre-

* The United States's total war dead is estimated at about 1,200,000.

viously colonized peoples, for example) or as an extant reality to be endorsed or challenged. The nation-state now enframes, for better or worse, the political actions and identities of the vast majority of men and women.

Nation-states can exist *on paper* before they exist *in fact*. Such is the story of the United States. For a *united* United States is a historic construction that most visibly comes into being as cause and consequence of American involvement in the Great War. Prior to the nationalist enthusiasms of that era, America was a loosely united federation with strong local and regional identities. The state's long arm did not yet reach everywhere, although a centralized federal government did exist (having grown apace under Lincoln in the Civil War era when one great wave of "nationalization" took place).

More important, perhaps, in the telling of this particular tale is the presence on American soil of millions of immigrants who flooded our shores in great waves throughout the late nineteenth and early twentieth centuries. Such "hyphenated Americans" were a source of concern to many "true Americans" and nationalizing politicians alike: Italian-American, Irish-American, German-American, Polish-American—who or what were they really? Could they be trusted to be loyal to their new home or had their hearts remained in the old? These questions kept various individuals and groups awake at night, fretting over the excessive infusion of alien and alienating people and philosophies into the American bloodstream. The First World War solved the problem. Here is part of that story—a dream and nightmare that the vast majority of women shared with the vast majority of men.

POPULAR NATIONALISM

Perhaps the place to begin is with the doctrine of popular nationalism as it took shape in Europe in the nineteenth century. Popular, or liberal, nationalism is generally associated with the hope that each nation of the world be able to form a state of its own. It is a principle of self-determination of peoples. But the advocacy of such nationality frequently elides into the less attractive notion that one nation (state) has claims to greatness or expansion that requires its dominating over other, lesser nation (states).

The best-known advocate of national self-determination in the nineteenth century, Joseph Mazzini, envisaged a world of distinct nation-states, each the locus of "the sentiment of love, the sense of fellowship which binds together all the sons of that territory. . . . The Country of all and for all."[21] His utopian vision was of an international federation of sovereign and equal nation-states working together in nonwarlike ways. Yet even within his attractive ideal of *nationality* lay the seeds of an intolerant *nationalism*. For that ideal turns on the celebration of "the people," the image of the many humming—in Mazzini's vision, they would rather be singing together than marching off to war—in unison to the same tune. When national unity becomes the highest good, the *point d'appui* of political and social life, it is all too easy to sacrifice other goods, including liberty, to that end.* For the advocate of strong nationality, too much stress on the individual gets translated as the promulgation of an anarchic and destructive *individualism*.† The idea of democracy as popular sovereignty is bound up historically with the idea of nationality—but how does democracy hold up under the strong aims of nationality transformed into the claims of nationalism?

The short answer is: rather shakily. At stake is not the rapacious nationalism that terrorized our century before and during the Second World War, the sort clung to by Benito Mussolini, for example. Nor am I pre-eminently concerned with Hitler's supranationalist dream of grand collectivities based on race. Both Mussolini and Hitler brought to their ultimate the deadly tendencies at work within earlier nationalist and racialist doctrines. Mussolini, for example, proclaimed that "the fascist conception of life stresses the importance of the State [which he always capitalized in his writing] and accepts the individual only in so far as his interests coincide with those of the State. . . . The fascist conception of the State is all-embracing; outside of it no human or spiritual values can exist, much less have value."[23] Imbued with this statist nationalism, a revitalized nation will go forth, from its aggressive

---

* Recall Aristotle's insistence, discussed in chapter 2, that, if a polity is geared for a single aim, that end is invariably war.

† Emile Durkheim endorsed a version of patriotism that is "internally oriented, fixing upon the tasks of the internal improvement of the society. . . . It prompts . . . nations . . . to collaborate towards the same end. [But] centrifugal [patriotism] prompts nations to encroach upon one another . . . they are put in a situation of conflict." This way is "aggressive, military"; the former is "scientific, artistic, and, in a word, basically pacific." As attractive as this distinction is, it is difficult to sustain in practice—or so a genealogy of the concept and its vicissitudes seems to suggest.[22]

energy, to reclaim that which is its own, to take from the weak, for the strong shall inherit the earth.

We tend to think of nazism in much the same way—a militant and narrow nationalism; certainly many Germans experienced it as such. The theory of nazism, however, included but went much beyond militant nationalism, codifying one of the most pernicious notions of twentieth-century thought: the insistence that blood or race is the biological basis for human collectivities. Hitler was quite clear in *Mein Kampf* that "we, as Aryans"—not Germans—"are ... able to imagine a State only to be the living organ of a nationality"—a transnationalism that respects neither state nor national frontiers. For the highest purpose of the state devoted to the interests of the *Volk* (the "people") is care for the preservation of "those racial primal elements which, supplying culture, create the beauty and dignity of a higher humanity."[24]* Hitler even went so far as to suggest that the word *nation* was hopelessly perverted through its links to nineteenth-century liberal ideals of nationhood of the sort embodied in Mazzini's life and work.†

Compared with the horrors of fascism, nazism, and Stalinism, the nationalism of the First World War era seems relatively benign, despite the carnage of the western front. Americans in particular see themselves in that era, and later, as a people driven to war by the nefarious deeds of others, or by the genuine needs of others for our assistance and protection. Our own nationalistic excesses get hidden within a story of a great and diverse people coming together only under extreme provocation to enter conflicts we had no real hand in "starting." We also look back from our current vantage point and find all our wars, with the exception of Vietnam, good, though some were better than others. The Second World War gets top billing as the "good war," but the First World War isn't far behind. We know very little about the wrenchings and excesses of our own moves to create a nation-state in the modern sense. But a part of that tale is one I aim to retell briefly,

* Contrast Franklin Delano Roosevelt's wartime proclamation of 3 February 1943, declaring, "Americanism is not, and never was, a matter of race or ancestry.... The principle on which the country was founded and by which it has been governed is that Americanism is a matter of the mind and heart."[25] This, at least, is the ideal.

† Hitler's supranationalism aims, as Hannah Arendt points out in *The Origins of Totalitarianism*, looked for a dominating superstructure that would destroy all homegrown national structures. For the body politic, the *only* home for a patriot, was too narrow and modest a compass for Hitler's supranationalism.

bringing to the forefront those features that have tended to be submerged or forgotten. We are no strangers to bellicist excess, and that "we" includes men and women, the educated and the uneducated, professionals and workers alike.

The time is the First World War. Europe—the Old World—has been at war since August 1914. Swept up in war's enthusiasms, three million men in England volunteered for the war before compulsory military service was introduced. In all combatant nations there was, at least initially, a passionate outburst of nationalistic zeal and an irresistible compulsion to portray the "enemy" in the most debased possible ways. Because the nationalism of this period was characterized by militarism, self-identification as a nation, writes Michael Howard, implied "almost by definition alienation from other communities and the most memorable incidents in the group-memory consisted in conflict with and triumph over other communities. France *was* Marengo, Austerlitz and Jena.... Britain *was* Trafalgar.... Russia *was* the triumph of 1812.... Could a Nation, in any true sense of the word, really be born without war?" Howard argues that when "the young men of Europe went out in 1914 to die in their millions, they did so for an ideal epitomized in the three words, God, King, and Country; and for those who recognized neither God nor King, *La Patrie* provided an adequate substitute for both. In 1914, in a historical moment of incandescent passion, the Nation almost in its entirety was fused with the State."[26]*

This incandescent passion swept (nearly) everyone up in its wake—young and old, male and female, working class and bourgeoisie. Nor were feminists exempt.† *The Suffragette*, the newspaper of the Women's Social and Political Union, was renamed *Britannia* and dedicated to

* It is worth noting that America emerged out of no such definitive victorious test against an external opposing force. Our most severe testing had been against ourselves in the Civil War, and the Spanish-American War didn't quite "make it" in the mytho-nationalist sense of a definitive and constitutive test of national manhood in the Hegelian mode. But the Spanish-American War did provide a base for what was to follow. It helped make notions of expansionism acceptable. Some demurred at the time, including the American philosopher William Graham Sumner who wrote in an essay: "Patriotism is being prostituted into a nervous intoxication which is fatal to an apprehension of truth.... The field for dogmatism in our day is not theology, it is political philosophy. 'Sovereignty' is the most abstract and metaphysical term in political philosophy. Nobody can define it. For this reason it exactly suits the purposes of the curbstone statesman."[27] Sumner's was a voice soon to be drowned out.

† During the First World War, some thirty-two national women's organizations can be identified, most of them mobilized *for* the war effort.

king and country. Emmeline Pankhurst, WSPU's redoubtable leader, made a terse and unambiguous announcement: "I who have been against the Government am now for it. Our country's war shall be our war."* George Bernard Shaw has described young women handing white feathers to all young men not in uniform. Sandra Gilbert details examples of the upbeat literature and war-fevered actions of many women writers and feminist activists in the period: Jessie Pope, in her "jingoistic 'The Call,' crie[d] "Who's for the trench/Are you, my laddie?" May Sinclair described "the ecstacy" of battle in *The Tree of Heaven*. Rose Macaulay, in her poem "Many Sisters to Many Brothers," expressed envy of the soldier's liberation from the dreariness of the home front: "Oh it's you that have the luck, out there in blood and muck." Concludes Gilbert: "In the words of women propagandists as well as in the deeds of feather-carrying girls, the classical Roman's noble *patria* seemed to have become a sinister, death-dealing *matria*."[28]

Caroline E. Playne details some of the excesses of "society at war." When the Women's Emergency Corps was created, women queued up in long lines waiting for the doors to open so they could sign up. Relief Committees were "everywhere being set up before the war was a week old" in England. The novelist Mrs. Humphrey Ward, in her book *England's Effort* (1916),† embraced classic rhetoric of the forging of national will against a foe construed as evil: "Then came the hammerblows that forged our will."[29] The war had many of the characteristics of a *crusade*, a war of a particular kind featuring a fight against a foe construed as thoroughly evil, a collective demonic entity to be crushed.

Towns and villages quivered with war excitement as people poised on the edge of intense, collective exhilaration. In *King Edward, the Kaiser and the War*, Edward Legge notes: "The harmony and pride of the civilian element . . . cannot be exaggerated. The wearer of the uniform is the idol, the hero of the day; just as the laggard is frowningly regarded with chilling indifference, often mingled with disdain. English women and girls have developed into ebullient, 'boiling' patriots, whose

---

* Emmeline Pankhurst and her daughter Christabel ran the WSPU with autocratic verve. Another daughter, Sylvia, split, going a more socialist and democratic direction than her mother and sister.

† Written at the request of Teddy Roosevelt, whom Mrs. Ward had met during her American visit of 1908, *England's Effort* was syndicated as a series of articles, "Letters to an American Friend." It is said to have hastened President Wilson's entry into the war.

value as recruiting agents cannot be overestimated."[30] In a debate in the House of Commons on 16 November 1916, Winston Churchill extolled women as being largely responsible for "unchecked and indiscriminate voluntary recruiting—enforced by every form of social pressure, equal almost to the power of compulsion of law."[31]

AMERICA BECOMES THE UNITED STATES

> Come, I will make the continent indissoluble,
> I will make the most splendid land the sun ever shone upon,
> I will make divine magnetic lands,
> With the love of comrades,
> With the life-long love of comrades.
>   —WALT WHITMAN, "For you o democracy"

Whitman's essentially pacific democratic vision of "inseparable cities" and an indissoluble continent is a tough dream to embody in and on the bodies of recalcitrant, diverse human material. We as a people did not come close to this dream until the era of the First World War, and then it seemed as much a nightmare as the stuff of a democratic utopia. Before the United States entered the war, various individuals and groups made pre-emptive strikes in its favor. An American journalist, Mrs. Mabel Potter Dadgett, was sent to Europe by her editor at the *Pictorial Review* to find out what the war meant "to the woman's cause," and reported in a book called *Women Wanted* that wartime adventure held enormous attractions. She got pretty close to the front and, at the end of 1916, concluded on a note of near epiphany: "I know of no more impressive place to be in the closing days of the year 1916 than here at the front of the terrible world war."[32] Terrible but wonderful, seductive and sublime, in Dadgett's telling of it.

In more muted civic rhetoric, the National American Woman Suffrage Association prepared itself for war in 1914 by proclaiming women's loyalty, patriotism, and professed readiness to serve in a variety of detailed capacities "in event of war." Thus NAWSA pledged itself to the country "and forthwith placed a plan of intensive service at the Government's command in view of the impending peril." Carrie Chapman Catt, president of NAWSA, repudiated charges of pacifism that had been leveled against suffragists:

We grant that we are in favor of peace; we grant that we have a large sympathy for the sufferings of humanity, but we also claim to be possessed of intelligence and knowledge and these have convinced us that there could be nothing more disastrous to the human race than *a peace at this time*, which would lead to greater suffering than a continuation of the war. Therefore, because we love peace and because we have large sympathy for human sufferings, *we are opposed to anything that will bring a peace which does not forever and forever make it impossible that such sufferings shall again be inflicted on the world.*[33]*

Women, Catt argued, are courageous, patriotic, and devoted to the service of the country—as proved by the fact that they were the *first* organized body "to formulate a definite line of action and present to the President and the Government a plan which would be followed by its more than 2,000,000 members provided hostilities went so far that war should be declared."[34] When war was declared President Wilson could, for the most part, count on the women. Among the departments of work the association declared its willingness to undertake were employment bureaus for women, increase of the food supply, elimination of waste, the Red Cross, and Americanization aimed both at integrating "eight millions of aliens" into the American way of life and emphasizing tolerance of others for them.† Alas, tolerance and integration proved to be antinomies rather than allies, as the often unsavory tale of the emergence of an American nation-state demonstrates.

The discourse of armed civic virtue in this era was the creation of nationalizing liberals, those who championed universal military training and conscription as an "effective homogenizing agent in what they regarded as a dangerously diverse society. Shared military service, one advocate colorfully argued, was the only way to 'yank the hyphen' out of Italian-Americans or Polish-Americans or other such imperfectly assimilated immigrants." The American playwright Augustus Thomas,

* There was a split in suffrage ranks on NAWSA's official stand on the war. Catt herself was also involved in the Woman's Peace Party.

† Blacks and black groups were divided over the war: for some, it was a white man's struggle; other blacks, however, saw in the war a "God-sent blessing," in the words of one Negro newspaper, "to earn white regard and advance the standing of the race by valiant wartime service." The Central Committee of Negro College Men petitioned Woodrow Wilson and the War Department to establish an officer's training camp for training "Colored officers for the Colored regiments in the New Federal Army."[35]

proclaiming France a "much finer democracy than our own . . . after a while . . . came to learn that that relationship [between social classes inviting fraternal compliance] had been acquired by men . . . working in fine equality in their military training."[36] In 1917, President Wilson committed himself to universal service, not just for defense in case of hostilities but to help mold a new nation.* Two years before the war resolution, Wilson had launched an attack against hyphenated Americans. "There are citizens of the United States, I blush to admit, born under other flags but welcomed under our generous naturalization laws to the full freedom and opportunity of America, who have poured the poison of disloyalty into the very arteries of our national life. . . . Such creatures of passion, disloyalty, and anarchy must be crushed out. . . . The hand of our power should close over them at once."[37] Not only did Wilson do nothing to stem the tide of xenophobic excess, he pandered to it. Congress responded in kind by passing legislation to check espionage and treason that vested the government with powers to censor the press, punish interference with activities of the armed services, including recruitment, and control the mails to prevent dissemination of allegedly treasonable material.†

An offensive to capture the minds of men and women got launched at the highest levels of the federal government. Under George Creel's audacious leadership of the Committee on Public Information, the nationalizers set to work, instituting, among other efforts, the "Four-Minute Men." Seventy-five thousand Four-Minute Men were drawn from local communities among applicants endorsed by at least "three prominent citizens—bankers, professional or business men . . . certified as to speaking prowess and safe political views, the men were turned loose to four-minute stints before any available audience to whip up

* Opponents of conscription argued that it threatened American ideals of individual liberty and voluntarism: these voices disproportionately came from the South and the West—from Speaker Champ Clark (D.) of Missouri who proclaimed that for Missourians "there is precious little difference between a conscript and a convict"; and Senator George W. Norris (R.) of Nebraska. It was a western woman, Montana's Jeannette Rankin (R.), the first woman to sit in the United States Congress, who voted no on the war resolution, 6 April 1917.

† David M. Kennedy's *Over Here* is chock full of the sordid details of these events. The columnist George Will's doubts about a "national community," which currently lead him to take Wilson to task for his "war corporatism," echo criticisms of radicals, such as Randolf Bourne, at the time.[38] Thus the political worm turns in ways that make ideologues of any stripe squirm. The picture, to serve their ends, shouldn't be so murkily out of focus.

115

enthusiasm for the war." The fact that America was composed of diverse nationalities is ironically evident in the fact that the seventy-five million copies of thirty pamphlets put out by Creel's CPI to explain the nation's war posture were printed in several languages. With the Four-Minute Men instructed to spread atrocity stories, citizens were encouraged to report to the Justice Department anyone who "spreads pessimistic stories . . . cries for peace, or belittles our efforts to win the war."[39]

Vigilantism against alleged slackers and traitors received public sanction. The lynching of Robert Prager, a young man who had been born in Germany (and had tried to enlist in the American navy only to be rejected for medical reasons), took place before a cheering crowd of five hundred near St. Louis. The mob's leaders were tried. The defense counsel called their offense "patriotic murder," and a jury returned a verdict of not guilty. Kennedy quotes a comment by the *Washington Post*: "In spite of excesses such as lynching, it is a healthful and wholesome awakening in the interior of the country."[40] Similarly, the *New York Times* deplored lynching but contributed to the public mood that invited it by attacking radicals (such as "IWW agitators") as "treasonable conspirators against the United States." John Dewey, the most important theorist of Progressive movement and pragmatism, rather than staunchly opposing intolerance became an apologist for the "overwrought state of public opinion," defining it as "not altogether unlovely. . . . The amusement aroused by the display is tinctured with affection as for all the riotous gambolings of youth."[41]

A vigilante organization, the American Protective League, received quasi-official status in its efforts to spy on neighbors, "fellow workers, office-mates, and suspicious characters of any time." Kennedy details the way members of the League opened mail, served as *agents provocateurs*, and led "slacker raids" against supposed draft evaders.[42] Other semiofficial and private organizations included the American Defense Society, the National Security League, the Home Defense League, the Liberty League, the Knights of Liberty, the American Rights League, the All-Allied Anti-German League, the Anti-Yellow Dog League, the American Anti-Anarchy Association, the Boy Spies of America, the Sedition Slammers, and the Terrible Threateners. A protest against the war staged in Boston, on 1 July 1917, organized by various radical organizations, was attacked by soldiers and sailors. A Socialist paper, *New York Call*, admittedly partisan, described the affair in this way: "Hundreds of Socialists were beaten and forced to kiss the flag on their knees. The headquarters of the Socialists were raided and wrecked. . . . Police were powerless or passive, seeming to sympathize with the enlisted men and the hundreds of rowdy volunteers who assisted them."[43]

The Supreme Court endorsed wartime assaults on dissenting opinion, weaving into the nation's legal fabric restrictions on freedom of speech that had been unknown before 1917.* America's school system was mobilized for the duration. Just as the schools and the army were nationalizing instruments in the creation of "France" (see chapter 2), so schools and the army became the great engines of nationalization in America. The one hundred thousand school districts into which America's school system was decentralized "lent themselves," in Kennedy's words, "to a kind of ideological guerrilla warfare"; hence "the struggle to control teaching about the war," to bring into being mobilized, armed civic virtue, "had to be waged in countless local actions, in communities scattered across the country." Plans for study for elementary school children were drawn up around the themes of "patriotism, heroism, and sacrifice." Atrocity stories became the fare of grammar school classrooms; and children were frightened with the notion that, unless Germany was defeated, its soldiers might come to America and treat Americans, including children, in disgusting ways. National organizations sprang up in blooming, feisty array, including the National Education Association, the Committee on Patriotism through Education of the National Security League, the National Industrial Conference Board, and the National Board for Historical Service, "a group of historians devoted to the 'progressive' or 'New History' belief that the study of the past should promote present-day social reform"—a reform that, they concluded, included war.† The approved course of study allowed for no ambiguities. The National Board for Historical Service "rejected one commissioned syllabus be-

---

* In *Schenck* v. *United States*, for example, Justice Oliver Wendell Holmes, Jr., enunciated the doctrine that "when a nation is at war many things that might be said in time of peace are such a hindrance to its effort that their utterance will not be endured."[44]

† Kennedy has a full discussion of the war for, or upon, twenty-two million young minds.[45]

cause it raised doubts about 'the positive values of nationalism.' " This syllabus was the work of J. Montgomery Gambrill, a professor of history at Teachers College, Columbia University. His mistake, according to the NBHS, was to stress reconciling nationalism with internationalism and to criticize too narrow conceptions of nationalism. According to the NBHS, the "untrained teacher" might be deflected from portraying the war "as a conflict between autocracy and liberal democracy," pure and simple, given Gambrill's subtleties.[46]

America's institutions of higher learning went for the war with relish, using the universities to contribute to the war cause as an embellishment of the "service ideal" that had been part and parcel of the emergence of the modern university in America during the Progressive era. Long exponents of social service, professors leaped with alacrity into a visible public role, putting their minds and their students at the behest of the nation-state. For example, on 6 February 1917, the general assembly of Columbia University "went definitely on record as desiring to place itself at the services of the City, State and Nation in whatever way it could be of greatest use at the present time." The women at Columbia "requested and secured the establishment of the University Committee on Women's Work as part of the mobilization of the university for war."* America's campuses were turned into military training camps for the War Department with the organization of the Students' Army Training Corps program in the fall of 1918. At 516 colleges and universities throughout the country, 140,000 male students were inducted into the U.S. Army and assumed the status student-soldiers.[47]

The dream of unity—Plato's dream, Rousseau's dream, Robespierre's dream, Hegel's dream—had, for Americans, become national. In the minds of the architects of national unity through armed civic virtue, the presence of millions of undigested immigrants was an ongoing irritant.† "Americanization" became the goal, the watchword; for some, the threat: one nation indivisible. To be sure, genuine regard for the welfare of immigrant groups lay at the base of much progressive sen-

---

* Although men far outnumbered women on university faculties in this as in subsequent eras, the numbers were less lopsidedly male-dominated then as now.

† In this connection, when driving, a few years ago, on a state highway near Richmond, Indiana, I needed no one to explain to me why the sign outside a small town there read: "New Germantown *or* Pershing."

timent, including that expressed by the National American Woman's Suffrage Association, which believed that separatism and heterogeneity were synonymous with inequality and marginality.

But progressive and liberal opinion proved particularly susceptible to the cry for unity because of its stress on popular sovereignty, on the "voice" of the "people," the notion that there must be some common civic glue to bind the nation as one. If that voice could neither speak nor understand the same language, literally as well as rhetorically, then no American dream of unity could be realized. A body politic cannot be dismembered, not on the nationalist view. The temptation to forge unity is great and invited figures from Woodrow Wilson down to trim the sails of free speech on the grounds that the war against dissent was a war against civic dismemberment, a war for great national aspirations and an opportunity to forge a community that might encompass the entire continent. With national "service" the byword, the hearts of dreamers of unity quickened when ten million men presented themselves to be registered for the army, on 5 September 1917, without "serious incident." One draftee in five was foreign-born but would be "alien" no longer, becoming an American among other Americans in armed service to his country.

Gruber speculates that the service ideal "might have led in a different direction had a distinction been made between service to society and service to the state."[48] But service to society seemed too disparate, too local, too parochial for architects of national unity who sought human mastery over the dangerous drift of human affairs. The only way to gain such mastery, it seemed, was to strengthen the powers of the state: hence John Dewey's talk of war's "social possibilities."[49] Randolf Bourne's bitter irony retains its force nearly fifty years after he left incomplete his essay on "The State":

War—or at least modern war waged by a democratic republic against a powerful enemy—seems to achieve for a nation almost all that the most inflamed political idealist could desire. Citizens are no longer indifferent to their Government, but each cell of the body politic is brimming with life and activity. . . . In a nation at war, every citizen identifies himself with the whole, and feels immensely strengthened in that identification.[50]*

* Bourne left his essay unfinished because he fell victim to the flu epidemic of 1919, his untimely demise robbing his generation of its most penetrating and courageous social critic.

119

Because our own identities are so entangled with those of the nation-state, the tendency, especially in a time of crisis, is for our spines to stiffen collectively and our voices to speak as one. When we perceive a threat to our honor, our security, or our way of life, most of us prefer not to hear discordance, and all too many among us construe dissent as disloyalty. This is the dark underside of the legacy of armed civic virtue. Some, however, have resisted, attempting to demobilize their words and their actions. I move next, as promised, to this alternative discourse.

4

The Attempt to Disarm
Civic Virtue

Happy the gentle:
they shall have the earth for their heritage. . . .
Happy the peacemakers:
they shall be called sons of God.
        —From the Sermon on the Mount, *The Jerusalem Bible*

I N THE BEGINNING . . . politics was war: the tradition of Sparta
and Rome, of Machiavellian city-states and autonomous republics on
the Rousseauian model. The role women play in this dominant narrative is that of Spartan mothers and civic cheerleaders, urging men
to behave like men, praising the heroes and condemning the cowardly.
Women are also official mourners, lamenting the destruction of the
war although the most horrendous possibility of all is defeat of the
city, not the deaths of particular individuals including their own husbands and sons.

But there is another discourse, one that coexists (to the extent that
it truly exists) in tension with civic virtue as armed and bellicose. It is
not a Biblical tradition as such, for Yahweh sometimes requires that
enemies of his chosen people be displaced and slain. The Hebrew
concept of peace, *Shalom*, speaks to a sense of well-being *within* walls

that signify security.* Alongside this dominant motif there is another, expressing the hope that spears shall be beaten into pruning hooks and swords into plowshares, a hope appropriated and extended in early Christian pacifism. Finding in the "paths of peace" the most natural as well as the most desirable way of being, Christian pioneers exalted a pacific ontology.† Violence must justify itself before the court of nonviolence.

The Christian savior is the "Prince of Peace," and the New Testament Jesus deconstructs the powerful metaphor of the warrior central to Old Testament narrative. He enjoins Peter to sheathe his sword; he devirilizes the image of manhood; he tells his followers to go as sheep among wolves offering their lives, if need be, but never taking the lives of others in violence. Men and women are required to examine their actions in light of injunctions against violence in word and deed. Jesus' teaching is unambiguous: "You have learned how it was said to our ancestors: *You must not kill,* and if anyone does kill he must answer for it before the court. But I say this to you: anyone who is angry with his brother will answer for it before the court" (Matt. 5:21).‡

That is the story—the Sunday School version, one might say. But it is incomplete. For what emerged as the dominant Christian teaching on war and peace was just-war doctrine, by no means a pacifist discourse. Something happened. A Thucydides or a Machiavelli might say: "The world happened—its realities forcing revision of pacifist niceties." That is an important part of the story. Involved as well are conflicting human aspirations and understandings of what is good, just, and necessary. Reflection on the genealogy of just-war discourse in the West also affords a sobering, sustained look at the abiding difficulty presented to human beings who attempt to live under the terms of a regimen that is ceaselessly pressured by the wider surround to succumb to its reigning ethos and instrumentalities: force, violence, and fear.

* *Yeru'shalom,* Jerusalem, is "the name of an impregnable fortress." When peace is exalted, it is often as a condition that follows a time of war, serving as postscript to a successful assault on Israel's foes.[1] There are, to be sure, discussions of limits to destruction in the Old Testament; but Yahweh's wrath is one of his central characteristics, and the warrior is a key figure in Old Testament discourse.

† More on "paths of peace" and the problems of defining peace in chapter 7.

‡ There is the Jesus who claims he has come to bring not peace "but a sword," meaning that his teachings will sprout strife and divisiveness.

The Christian Conundrum:
From Pacifists to Reluctant Warriors

Look with the eyes of the newborn child upon the men and women who are dying of hunger while enormous sums are being spent on weapons. Look upon the unspeakable sorrow of parents witnessing the agony of their children, imploring them for that bread which they have not got but which could be obtained with even a tiny part of the sums poured out on sophisticated means of destruction, which make ever more threatening the clouds gathering on the horizon of humanity.
—Pope JOHN PAUL II

Although the pacifist and just-war streams in Christianity parted historic company, they remain genealogically related. Both put violence on trial, placing the burden of proof on those who take up arms rather than on those who refuse. Both evaluate social life *from the standpoint* of the suffering, potential or actual victims, rather than of the triumphant, the victors. We are enjoined to "look with the eyes" of the vulnerable, the tiny, the suffering, making plain to the believer, and to all other persons of good will, that contemporary Christians should refuse to take up sides with those who erect the crosses of history but, instead, should stand with those who are crucified under the terms of unjust orders.[2]*

Two scholars, two identical conclusions:

Prior to the advent of Christianity, there is no record of anyone suffering death for a refusal of military service.[3]

There is no known instance of conscientious objection to participation in war or of the advocacy of such objection before the Christian era, and until roughly the last one hundred and fifty years pacifism in the West was confined to those who stood inside the Christian tradition.[4]

The New Testament backdrop went like this: although Jesus didn't

---

* Just-war thinking offers, with reference to social arrangements, a vantage point from which its adherents frequently evaluate, and find wanting, what the world calls "peace": thus, John Paul II characterizes our current armed-to-the-teeth peace as the continuation of war by other means, as not peace but one of the forms war takes in the modern world—a subtle and important point, which I pursue later in this chapter.

offer rounded-out views on war fighting, he refused to countenance military force to achieve his aims or to usher in the peace of the kingdom of God. In the context of his age, he was thus positioning himself against Rome and, as well, against the "warlike Messianism" of the nationalist Zealots.[5] What is less certain is whether Jesus was interested at all in "politics" or even in what that word might mean within a horizon set by the Gospels.* Jesus resisted the devil's temptations of power and earthly kingdoms but endorsed rendering unto Caesar that which is Caesar's; however, what was Caesar's didn't count for much in his eyes. What was *not* Caesar's was definition of life's ultimate purpose and each human being's vocation and calling. The politics of Jesus' time presents an unsavory picture of a bloated imperial hegemon holding sway over subject peoples and offering ample space for power machinations of the greedy and ambitious. Early Christians courted martyrdom by refusing to worship the Roman emperor as a divinity, an act that stripped the empire of one of its most dubious pretensions.

When Jesus blessed the peacemakers and chastised the powerful, he also offered up a challenge to the terms of the *Pax Romana*, for the emperors had gotten into the habit of pronouncing themselves the "Sons of God" given their duty to uphold the imperial peace. Dubbing his own humble peacemakers the "children of God," Jesus held up a concept of peace at odds with the notion that peace and unruffled order maintained by armies of occupation were one and the same. Even as he turned on their head classical scales of martial virtue and valor, Jesus redescribed the deity. The Old Testament God of Justice gave way to the merciful Father: "Be ye therefore merciful as your Father also is merciful" (Luke 6:36). The lives of the downtrodden and the outcast reverberated to new-found dignity as men and women took in the heady insistence that the last shall be made first.† Interrogating reigning notions of "good" and "bad," of "clean" and "un-

---

* The Christian revolution was, if anything, a counterpolitics, the articulation of that freedom *from* politics which Hannah Arendt has called one of the most significant features of the Christian heritage, and which has been reappropriated again and again in historic struggles against authoritarian and, in our century, totalitarian regimes.

† In Jesus' teaching, this is not a glorification of suffering and an advertisement for future turning of the tables but a recognition that his message overturned received expectations and presaged radical social uncertainty.

clean," Jesus deflated the terms of law and authority with parables that defied both.*

Aiming to liberate human beings from the need to have victims and to sacralize the victims that they have, Jesus assaulted the edifice of the "false transcendence" of violence: that is, structures of experience that require constituting enemies and slaying them and setting up scapegoats. By exposing the violent origins and making visible the distorted passions required to create and to sustain destructive structures of social life, Jesus not only terrified his opponents but unhinged his followers as well who sometimes professed their befuddlement about what, exactly, was going on.†

How did this message play in Palestine and throughout the ancient world? Once Christians gained a toehold, they extirpated the barbarism of the Roman arena; forbade infanticide and the abandonment of infants—which meant sparing more females than males this fate, the former being more likely to suffer it; and stripped the warrior of his honored status. The Council of Nicaea, in the fourth century, declared that the Christian who had abandoned warrioring only to return to it did so as a dog to its vomit. Christian writers in the first several centuries A.D. tended to conflate wartime killing with murder and labeled war itself as iniquitous, mad.‡ Near the end of the third century, a young man from Numidia became the patron saint of conscientious objectors. Called up for military service (his father had been a Roman soldier), Maximilianus told the Roman proconsul that, as a Christian, "I cannot serve as a soldier; I cannot do evil." He was executed for his persistence.[7]

---

* For example: a laborer hired at the "eleventh hour" gets the same reward as one who has worked all day in the heat; a feast is thrown for the prodigal son upon his return, as the son who has doggedly done his duty does a slow burn; the woman who ignores household tasks in order to listen to Jesus' teaching is praised, to the distress of her bustling sister (the story of Mary and Martha). I am not hinting that Jesus is one of the early demystifiers of the "feminine mystique"; simply, his priorities shook up given pieties, roles, and expectations. Only if one gets a glimmer of the radical nature of his teaching can one make sense of the expectation of early Christians that they had heard the "good news" and been liberated from the world's ways.

† The unsettledness Jesus stirred up when he released people from bondage to demons of their own devising suggests that radical moral critics of established orders, particularly peacemakers, are almost certain to trail in their wake confusion, anger, havoc, aroused vigilance. A mistake frequently made is to assume that making peace is peaceful, placid, orderly, quiet.[6]

‡ As well, priests forswore use of their sexual instrument. The phallus as an instrument of power and domination is fretted about in the work and lives of such historically and culturally separated men as Augustine and Mahatma Gandhi.

This sort of principled resistance to public power was something new under the sun, opening up a range of options, duties, responsibilities, dilemmas, and reassessments not available in classical antiquity, bearing implications for men and women alike.

Intimations of altered social identities of, and for, male and female also become visible.* The model for Christian love, *agape*, was the mother's unconditional love for her child, marking a feminization of Christian ideals of fellowship and community. The qualities most often associated with maternal imperatives were urged above all. In the vocation of martyrdom, women achieved equality with men in an arena of torture and terror, making possible a parity in the purposes to which men and women put their bodies, a parity unthinkable within the frame of classical armed civic virtue.

One early female martyr, a young catechumen in her twenties, the mother of an infant son, was arrested, tried, imprisoned, and put to death in A.D. 203. Able to read and write, she left behind a vivid, direct narrative of her trial and testing, which has come down as the *Passio SS. Perpetuae et Felicitatis*. The modern reader is startled by Perpetua's extraordinary bravery and her sure and certain insistence on her own agency, a gentle, unswerving determination to be true to the name she chose for herself rather than give way to her devoted father's entreaties to recant and be saved. She "grieved" for her father's despair but would follow "the authority of my name": "I am a Christian." Perpetua recounted several dreams bursting with symbolism and allegorical meaning. For her they were visions in which she *won* her combat and prevailed against terrible foes, thus preparing her for the arena: for she "knew I should have to fight not against wild beasts but against the Fiend, but I knew the victory would be mine" (that is, the temptation to recant, prompted by the Fiend, is a worse fear by far than having her body torn apart by desperate animals in the arena).[8]

Perpetua's agony and triumph were "named" by her as an act of free-willing that put her at odds with the comforting and comfortable world of her familiar surround. Her moment of individual testing and personal transcendence is presented with a lyrical vitality that signifies the author's individuality and her power or, better, points to the power

---

* While the story of traditional patriarchal reaffirmations inside the institutional church is not my focus here, the reader should be aware of the incompleteness of any moral revolution with reference to prior understandings.

of early Christianity to produce extraordinary "selves," to reveal capacities of men and women—young and old, drawn from all social strata—to endure and to prevail. I unfold this at length because one of the perennial vexations of those who oppose war is that they cannot seem to enlist in this effort the same soaring energies unleashed in time of war. The saga of Christian martyrdom suggests that dramas of certain sacrifice for one's beliefs, those beliefs having become *oneself*, may function in analogous ways. Strength in and through undeserved suffering is a consistent benchmark of pacifist thought and action.

But the "soldier question" wouldn't go away. By the fifth century, arms bearing, with restrictions, made a comeback as a Christian ethic of war was adumbrated. This partial reversal wasn't as total and unseemly as it might have been, given the fact that Christian discourse had never been entirely "war-free." Martial metaphors were frequently deployed by Christian writers. The Book of Revelations is a phantasmagoric tale of apocalyptic struggle against the Beast; and, as well, warfare got absorbed into Christianity as a foundational text when Christian leaders accepted the Old Testament as a divinely inspired portion of Holy Scripture.*

One way to consider the shift from pacifism to conditional acceptance of collective violence is to see it less as a definitive renunciation of pacifism, more as a discursive tremor that resituated pacifism as a (partly) submerged doctrine. Pacifism took on a particular spatial location, within the bodies and communities of monks, who became prime transmitters of the nonmilitary tradition. Women figured in pacific representations as Christian mothers in the image of the Madonna, but this was a less decisive shift than that made by priests who renounced their sexuality in becoming pacifists. The fighter is reborn in the image of a Just Warrior who takes up arms reluctantly and *only* if he must to prevent a greater wrong or to protect the innocent from certain harm. His tragic task is made necessary because the dream and hope of peace on earth has been indefinitely postponed.

In part, a tale of the "return of the repressed"; in part, a response to the barbarian invasions which, with the sacking of Rome in 410,

* A "revelation" is, in Greek usage, an "apocalypse," a distinct literary form larded with rich symbolism and hidden messages. Modern apocalyptic writers are more didactic, less allusive than is the author, or authors, of the New Testament contribution to our collective fears and tremblings.

signaled an end to the universal dream of the *Pax Romana*; in part, the anticipated (by us) alterations required when a charismatic movement decides to stay around for the long haul and become embodied in institutional forms: the supercession of pacifism by just war for some believers was, and remains, a historic tragedy of immense proportions; for others, however, it is a prudent and responsible response by human beings living in a dangerous, occasionally disordered world.

Ironically, the earthly order the Christians had reviled and been persecuted by was also the order they had come to *assume*, a structure of law and administration they expected. When Rome fell, the collective policeman was demobilized, and all the people he alternately protected and bullied trembled at this turn of events.* Peering into the abyss that opened and swallowed Rome, Christians shared in the profound disturbances of all late antique peoples. One could either retreat into oneself and await the end—which threatened to come rather soon— or devise some way to turn the teachings of the Prince of Peace into a doctrine that both condemned and condoned collective violence depending upon the circumstance, the situation, and the ends sought.

Just War, Holy War, and the Witness of Peace

AUGUSTINE IN HIS TIME

> Those wars are now past history; and yet the misery of these evils is not yet ended. For although there has been, and still is, no lack of enemies among foreign nations, against whom wars have always been waged, and are still being waged, yet the very extent of the Empire has given rise to wars of a worse kind, namely, social and civil wars, by which mankind is more lamentably disquieted either when fighting is going on in the hope of bringing hostilities eventually to a peaceful end, or when there are fears that hostilities will break out again.
> —SAINT AUGUSTINE, *The City of God*

Located as the father of just-war teaching, occupying discursive space similar to that taken up by Machiavelli or Hobbes inside the realist

---

* A consistent criticism of individuals and groups who prejudge all war as immoral is that they frequently receive the protection of governments in exchange for political passivity.

1

(1) Joan of Arc as portrayed by Ingrid Bergman in the 1949 film.
Photograph used by permission of King World Productions.

As Joan of Arc—in her real as in her imagined person—is the prototype of honor, bravery, and sacrifice, so women nearer our own time have flouted society's expectations and decked themselves out as men to subscribe to the warrior code: thus, Kady Brownell in the Civil War (2); and thus, Calamity Jane, herself a symbol—with gunfighters and frontiersmen—of the "wild" West (3). While today, not as isolated heroines but in numbers, women are joining the Militant Many as the almost equals (but for the rule against their engaging in direct combat) of men—as exemplified by the 1980 film *Private Benjamin* (4) and by this woman marine (5) being trained in hand-to-hand combat—and so take on not only the virtues of the warrior tradition but its hard realities as well.

(2) "Kady Brownell," engraved by G. E. Perine & Co., N.Y. From Frank Moore, *Women of the War* (S. S. Scranton & Co., 1866). (3) Courtesy of the Bettmann Archive. (4) Goldie Hawn in *Private Benjamin*, © 1980 Warner Bros., Inc. All rights reserved. (5) Photo by Susan Meiselas, Magnum Photos.

6

7

A thorny complex of images attaches to the Beau
tiful Soul and the Just Warrior. Three centurie
before Rambo (6), exemplifying the heroic lon
at war with the system, a real Hannah Dusto
massacred her Indian captors. Represented as
true heroine, she stands in this 1874 statue (
clutching the scalps of her ten Indian victim
Quite other is this saintly Beautiful Soul "befor
the battle" (8) and the mother sending off her litt
son to be a drummer boy in the Civil War (9)–
both women enthralled by the vision of nob
sacrifice; both, at least temporarily, blind to th
dirt, the turmoil, the terror, the inhumanity
war.

(6) Sylvester Stallone as Rambo. From *Rambo First Bloo
Part II.* © 1985 Anabasis Investments N.V. Carol
Service, Inc. (7) Photograph of the statue of Hann
Duston courtesy of the New Hampshire State Gover
ment. Photo by Saunderson. (8) "Before the Battle," e
graved by G. E. Perine & Co., N.Y. From Frank Moor
Women of the War (S. S. Scranton & Co., 1866). (9) "
Mother's Sacrifice," engraved by G. E. Perine & C
N.Y. From Frank Moore, *Women of the War* (S.
Scranton & Co., 1866).

The decades since the Second World War have seen the ranks of anti-war women protesters grow even as many soldiers who fought in Vietnam came to see their fate as tragic and themselves as betrayed. Thus, we have moved *from* this strong female figure commemorating those who fought and died in the Second World War (10), and the Gold Star Mother who lost five sons to it (13, *overleaf*), *to* this veteran of the Vietnam War throwing his crutch away in protest in 1971 (11) and *to* this pile of medals, won in that war, discarded by his comrades in front of the Capitol (12), and *to* the women of Greenham Common protesting missile deployment in England in 1982 (14, *overleaf*)—not to negate those past exemplars of valor and sacrifice but to place alongside them women and men who dare to say no to war and to offer instead new identities and possibilities.

(10) Pacific World War II Memorial. Courtesy of American Battle Monuments Commission, Washington, D.C. (11) (12) (13) Courtesy of AP/Wide World Photos, Inc. (14) Photo by Chris Steele-Perkins, Magnum Photos.

13

14

frame, Augustine presents problems for those who like their philosophies neat.* Of his 117 books (written in the late fourth and early fifth centuries), none is explicitly a work of political theory or moral philosophy in a systematic sense. He shares few of the presumptions that guide either the classical realists or, for that matter, orthodox just-war thinkers. For starters, for Augustine the launching pad for sustained reflection is *not* the state but the *saeculum*: that is, human existence as experienced and inherited since the Fall. He rivets a relentlessly skeptical spotlight on *all* human action.

Perhaps I should explain my rather extended treatment of Augustine in a work that is not primarily a scholarly tome in political philosophy. The power of his prose tempts one to linger, but there is a more important reason: namely, Augustine's way of working things through is an exemplary alternative to the abstract, rationalist castings of much contemporary political and moral thought.

Take a look at history, he suggests: observe, first, the unavoidable miseries—illness, death, famine, flood; then witness in sorrow the ways man has devised to compound these torments with his own inanities. If he is overtaken by a lust for dominion, a human being will find pleasure in "every kind of cruel indulgence"—Augustine has in mind the Roman games; and anyone "who disapproves of this kind of happiness" will rank "as a public enemy" and be "hustled out of hearing by the freedom-loving majority."[9]

Forms of "revolting injustice," catalogued by Augustine, include subjugation of the lower orders by the powerful, and the degraded mode in which the subjugated fight back; laws that disallow daughters to become heirs ("I cannot quote, or even image, a more inequitable law"); excessive concern for the opinions of others, exemplified in deformed codes of virtuous conduct for Roman women (for example, the code that sets up an expectation that any woman violated in time of war, having been shamed, will do the honorable thing by killing herself)†; perversion of marriage, the family, and social life by "the glory of possession" which makes a man feel "that his power was

---

* Some cavil at placing him inside just-war doctrine; he is more properly, they argue, a "Christian realist."

† Women violated against their will are "not to be ashamed of being alive, since they have no possible reason for being ashamed at having sinned," Augustine declares, rejecting female honor, Roman style.

diminished if he were obliged to share it with any living associate."

Augustine anticipates modern critical strategies when he strips "off the deceptive veils," removes "the whitewash of illusion," and subjects "the facts to a strict inspection." Take away the screens of "the splendid titles of 'honour' and 'victory' "—really "senseless notions"— and what one finds are "crimes . . . great evils to vex and exhaust the whole human race." Rome was conquered by her own lust to dominate as she triumphed over others: "Think of all the battles fought, all the blood poured out, so that almost all the nations of Italy, by whose help the Roman Empire wielded that overwhelming power, should be subjugated as if they were barbarous savages." The only law for Roman conquerors was that of vengeance—even when peace and victory were proclaimed.

"Peace and War had a competition in cruelty; and Peace won the prize." Augustine shifts to biting irony where the follies and fancies of Rome are concerned: "For the men whom War cut down were bearing arms; Peace slaughtered the defenceless." Let us not be taken in by "empty bombast" that blunts "the edge of our critical faculties . . . by high-sounding words like 'peoples,' 'realms,' 'provinces.' " A kingdom without justice (most kingdoms most of the time) is a criminal gang on a large scale. Augustine repeats the story of the rejoinder given by a captured pirate to Alexander the Great when Alexander queried him about his idea in infesting the sea. "And the pirate answered, with uninhibited insolence, 'The same as yours, in infesting the earth! But because I do it with a tiny craft, I'm called a pirate: because you have a mighty navy, you're called an emperor.' "

The earthly city is never free from the dangers of bloodshed, sedition, and civil war. Human beings are permanently estranged—from one another, from completeness and unity, from the attainment of permanent peace or justice or control over themselves, others, or human events. A human being cannot even be certain of "his own conduct on the morrow," let alone specify and adjudicate that of others in ways he foreordains. In this world of discontinuities and profound yearnings, of sometimes terrible necessities even "in his own actions," the best a human being can do is to maintain an order that approximates justice, strives to prevent the worst from happening, resists the seductive lure of imperial grandiosity. Think of the cost of "this achievement," Augustine reminds those who extol the universality of empire: "Consider the scale of those wars, with all that slaughter of human beings, all the human blood that was shed!" If you reflect on this, you will wage only just wars even as you lament the fact that they sometimes must be waged given the "injustice of the opposing side."

His acceptance of "just war" is reluctant. In detailing the abundant miseries of wars of all kinds, Augustine declares that he cannot "possibly be adequate to the theme," and "there would be no end to this protracted discussion," were he even to attempt to describe "the many and multifarious disasters." That the "wise man ... will wage just wars" is cast by Augustine in the form of an impersonal "they": "But ... *they* say." His grudging endorsement of the lesser evil enters his discourse as the promulgation of a collective Other. He, in turn, proffers it as the best of a bad deal. As to war's end or aim: there can be no talk of power, or civic *libertas*, or sovereignty, or national interest but only of *peace*. From the tigress pouring over her cubs, to human fellowship in all its forms, peace, even a perverted peace unworthy of the name, is the ultimate aim of conflict.

Augustine believes he has uncovered the lowest common denominators of human existence in the *saeculum*: a need for social life, hence for peace and order; a divided will easily traduced by a lust to dominate and to possess; a world of insoluble estrangements, perils, and shortcomings. *That* is the world we inhabit, reflects Augustine; and it is a world he evokes in densely textured description, through scathing indictments of folly and false pride, in passages that soar with the joys of friendship and resonate to the hope that estrangements will be healed and human incompleteness made full in the City of God.

Women figure centrally in Augustine's world: they are subject to the same yearnings and seductions as men; they can will, and choose, the way of *cupiditas* or *caritas*; they can acquiesce in the disorder of earthly dominion or share the fellowship of the Christian community and long for the Heavenly Jerusalem. Augustine's household, unlike Machiavelli's cordoned-off private sphere, is the "beginning or element of the city," and domestic peace bears an inner relation to civic peace. The household and the city do not diverge in kind; rather, aspects of the wider civic surround "press" human beings to conform.

The integrity or disintegration of the parts lofts upward to become a feature of the whole, an aspect of the city. Men and women are not divided into civic and noncivic beings. They are gifted with fellow feeling, vexed by desire, given to wickedness (by choice, not nature), and united by love. Augustine's definition of "a people" is of a group of human beings "bound together by common agreement as to the objects of love."[10]

A vast distance separates Augustine's discourse from later just-war doctrine—particularly in its modern refinements that stress legalities, rights, and the power of human reason. Although Augustine puts violence on trial, seeking to limit its occurrence and deflate its glory, and deploys the language of *just* and *unjust* to characterize collective conflict—lifted from the world of the City of God where they are embedded features of the landscape he so energetically, even relentlessly, evokes—such terms lose their evocative force. To understand politics, he insists, one must be attuned to what human beings love in common, what they fear, what they are subjected to, and what they can anticipate. The force of this extended Augustinian moment will emerge with greater clarity when I consider modern just-war discourse (see pages 149–59).

THE MIDDLE AGES AND THE CRUSADES

Those centuries tagged as "medieval" and dominated by a universalizing Christianity should, presumably, offer evidence of the way just-war teaching got embodied in social and political forms. But the picture is murky at best. The centuries A.D. 500–1000 are a time of ubiquitous, ceaseless conflict in what was to become Europe. Compounded from ancient survivals, Germanic codes, and Christian ideals and experience of community, medieval structures of "lived life" ranged from saintly ecstasies and mystical visions to sordid brutalities.

Popular campaigns, in which women participated, promoted the Peace of God and the Truce of God in the eleventh century. These were attempts to limit the days on which fighting could take place, and restrict those involved in conflict by exempting from assault and destruction clerics, monks, nuns, women, pilgrims, merchants, peasants, visitors to councils, churches and their surrounding grounds, cemeteries and cloisters, the lands of the clergy, shepherds and their flocks, agricultural animals, wagons in the fields and olive trees.[11] Feudal lords and knights were required to take oaths pledging themselves to these rather rigorous limitations on plunder and pillage. Here we find the Church responding to, and sanctioning, a popular outcry to limit violence.

The medieval warrior class consisted of a few, like the Japanese *samurai*, who devoted a lifetime to training for fighting and fighting. But the vast majority of the population was uninvolved, as warfare took the form of small skirmishes, raids, and encounters that sometimes added up to a battle. A prolonged siege might work hardship on women, children, and other noncombatants; but, for the most part, war was an unexceptional fact of life, chronic but limited, the vocation of an aristocratic minority. Sacralized, the sword was hallowed when wielded by a Christian knight.

Women are represented variously in this scheme of things. They serve as witnesses to male combat and valor in ritualized tournaments, turning such events into an occasion for sexual display and rivalry. Chivalry linked men to each other in the traditional ways of men-at-arms, through the bonding of shared risk and training, meals, and quarters. Christian knights seem not to have tormented themselves overmuch concerning the justice of their cause; instead, they acted in and through identities cemented by oaths of liege loyalty, sanctified and ritualized by religious belief.

Although women could be fief holders, they were not trained as knights or thought capable of defending themselves through force of arms. Some women did act as war leaders in their capacities as fief holders in their own right or standing in for absent husbands. The tradition of courtly love, which took root and flourished in the twelfth century, represented the lady as the obsessive object of male desirings, an obsession to which he submitted willingly, thereby putting himself under her power. Love and worship of her made him a better, even a purer man—or so the verse of the troubadours proclaims, for she permitted his adoration without yielding to him in the flesh. These complex negotiations fed visions of the "true woman" as a lady of aristocratic refinement, noble sentiments, a patron of the arts, and benevolent mistress of the manor.

That medieval religiosity was imbued with martial values at odds with just-war deflections is most evident in the emergence of the Holy War, or Crusade. The cause of peace—a desperate longing in Augustine—engrafted onto subsequent bellicist eruptions, reappropriates the Roman imperial ideal: a dream of order through conquest. Thus Urban II, at the Council of Clermont in 1095, inaugurated the Crusades by proclaiming, "Christ commands" that "this vile race" be exterminated "from the lands of our brethren": "Oh, what a disgrace if a race so despised, degenerate, and slave of the demons, should thus conquer a people fortified with faith in omnipotent God and resplendent with the name of Christ! . . . Let those who have hitherto been robbers now become soldiers of Christ."[12]* *Deus volt*—God wills it.

Despite crusading excesses and enthusiasms, prohibitions and restraints promulgated by the Church against violence, coexisting uneasily with a highly agonal warrior code, eventually passed into the practices of European nation-states and armies. The slow emergence of independent states also presented a sharper cleavage between times of war and peace, between battle fronts and home fronts, than had been characteristic of medieval life. Despite their engagement with practices that involved violence—hunting and trapping, for example—women were more and more construed as a collective, pacific alternative, as the dominant male chivalric warrior gave way to the soldier-citizen of the state. The force of these understandings, as they coagulated in years that mark the transition from late Middle Ages to Renaissance, is such that Christine de Pizan—a writer whose work challenged received notions of women, their faculties, and their moral status, herself the author of a treatise on the martial arts—took up (in her allegory, *The Book of the City of Ladies*) a position best described by the name she ascribed to one of the three founding goddesses for the special city reserved to women who had made noteworthy contributions to society: Rectitude, whose partners in founding are Reason and Justice.

Noteworthy is the falling away of any nastiness, any dilemma that might implicate the Wise Princess in the perennial dilemma of "dirty hands." Wars are dramatic by their absence in the City of Ladies (they *are* ladies, after all). Christine repeats the story of the Amazons, noting "delight in the vocation of arms," and their merciless policy of ven-

---

* Urban was referring, of course, to Muslims. *Plus ça change. . . .*

geance, in an exchange with Reason aimed at remembering women whose skills demonstrated that they were "fit for all tasks."[13]* Christine does not endorse the Amazonian way; indeed, she specifically rejects conflating the vocations and virtues of men and women. Accepting Amazonian escapades as historic truth, Christine nonetheless denies this narrative of female ferocity and martial vigor any contemporary clout. Her own vision of the "lady" and the Wise Princess bar such appropriations. Fashioning herself in part as a work of art, an aesthetic that distanced her from the structure of organized violence, Christine de Pizan anticipates, as well, later humanists, liberals, and rationalists who shared her fondness for Discretion, Reason, and Justice and found war irrational, stupid, wasteful, and atavistic.†

In a second work, *The Book of Three Virtues*, Christine plays on "Mirrors of Princes," advice to the Christian prince on how to be both just and resolute, firm and compassionate. She describes "The Life-Style of the Wise Princess, According to the Admonitions of Prudence." This lady, "who has been given some responsibility for government," is pictured as kindly, courteous, devoted and devout, an alms giver, a merciful hearer of requests, a presider over councils, weighing all proposals carefully, listening to everyone, and finally, after deliberation, noting that "which seems to her the wisest and most valid opinion." She appoints "competent gentlemen to be her advisers," entertains modestly, receives visitors, takes up handiwork, and hears vespers at the end of her busy day.[15]

THE PROTESTANT NATION-STATE

Martin Luther's self-understanding was that of an Augustinian wiped clean of the taint of Thomist Aristotelianism and papist perversions of the Word. Luther reworked Augustine's just-war teachings, accepting the view that the aim of a just war is peace, and seeing war as a last resort, a defense of a way of life against possible destruction. Temporal government, the rule of the princes, must have the power

---

* Exemplary ancient warrior queens—Zenobia, Artemsia—and medieval exemplars—Lilia, Queen Fredegund—are also recalled. Some bore arms themselves; others exhorted vigorously.

† Natalie Zemon Davis has argued that Christine finally could not decide "whether women were better than men or the same as men in regard to their virtue and their potentiality for aggressive or violent behavior."[14]

both to fight just wars and to suppress insurrection and disobedience. With other just-war thinkers, Luther's teaching is tilted to extant, officially legitimated units of rule. War is one of the police powers of the prince, and the good that pre-eminently defines the temporal kingdom is *order*. Luther's vicious condemnation of the Peasant Revolt, a series of uprisings between 1524 and 1525 by peasants pressed by economic change and anxious to recover lost feudal liberties, is an extreme example of just-war suspicions that internal insurgence most often presages greater bloodshed and injustice than it does a more just order.*

Luther's state is a political regime unleashed from the institutional constraints of the medieval church: the pope has no authority over lawfully established princes. Luther prepares the way for the political theology that underlies the emergence of the nation-state. Its full-blown dimensions become more visible in seventeenth-century calls for holy wars, providentially enjoined so that tyranny might be banished and the True Godhead worshiped. Centuries scarred by religious wars are one result of the turn, by militants like Oliver Cromwell, toward a holy-war ethos espoused in the sure certainty that the party of Liberty had a divine right to "execute judgement upon the Heathen . . . upon the King and his wicked Adherents."[17]

Following the excesses of Europe's religious wars, the crusading ethos does not disappear; it regroups, taking shape as the popular bellicism and militarism of the nineteenth century, feeding notions of sovereignty as a secular mimesis of God as penultimate Law Giver whose commandments must be obeyed and whose power to judge is absolute. Similarly, the triumphant state cannot be resisted, nor its will thwarted. Churches having been disarmed in their relation to the state—that is, hedged in by a *cordon sanitaire* that muted their potential political force or, more decisively, constituted churches as arms *of the state* (in England and Prussia, for example)—the likelihood that war's eruption would become the occasion for a crusade, a total struggle of

---

* At present, those indebted to the just-war tradition are prepared to accept "recourse to armed struggle . . . as a last resort to put an end to an obvious and prolonged tyranny which is gravely damaging the fundamental rights of individuals and the common good," but only after all other avenues have been exhausted. This is faithful to the original spirit of just-war and pacifist thinking: that violence must justify itself, with the burden of proof falling on those who engage in it, not on those who refrain.[16]

good versus evil, and no mere just (hence limited) war, was virtually guaranteed.

Thus the Bishop of London in 1914 exhorted his youthful countrymen:

Kill Germans—to kill them, not for the sake of killing, but to save the world, to kill the good as well as the bad, to kill the young men as well as the old, to kill those who have shewn kindness to our wounded as well as those fiends who crucified the Canadian Sergeant [referring to a widely circulated propaganda myth]. . . . As I have said a thousand times, I look upon it as a war for purity, I look upon everyone who dies in it as a martyr.[18]

The National Socialists, a generation later, could count on a pacified state church not to make trouble—and it did not, as churches were decked out in swastika banners and pastors cried, "Heil Hitler."*

The discourse of armed civic virtue and the ferocious excesses of holy war pursued for the state rather than for God came together in a way that fused undisguised *machtpolitik* with assumptions of state divinity, canalizing popular sentiment into collective enthusiasm for the "state ideal" as the self-identity and definition of "a people"; integrating whole populations into mobilized political patterns; producing young men prepared to make the "supreme sacrifice" and young women prepared to see to it that they did. The state devised education aimed explicitly at creating "generations physically fit for and psychologically attuned to war."[20]

Whether fighting a classic, defensive, just war or engaged in an orgy of self-righteous destruction in a "holy war," young men in the modern West have been ongoingly reinscribed with the identity of the Just Warrior. Again, the First World War represents an apogee in the way Western nations sanctified their war effort, turning it toward the crusading extreme. A few feminists of this era even managed to see in the orgy of destruction necessary punishment against those nations that

* One recent study indicates that the voting record "shows clearly" that Catholic areas gave less support to the Nazis than did Protestant areas, and "the Protestant churches as a whole were less helpful to Jews than was the Catholic church." The Confessing church, which broke with the state church was, of course, an important exception to the rule of compliance.[19] But none of the organized churches comes off particularly well.

had most harshly subjected women. The Countess of Warwick, for example, having swallowed whole British anti-Hun propaganda, concluded that the reason the "Boche" was so ruthless and must be defeated by the just was "the fact that in their country women are kept more in the background than in the country of any other great Power." On the other hand, in France, an ally she loved, Warwick found that "woman rules. . . . Feminism is one of the strongest forces in France." The "women of the world" rightly turned against Germany "when the manner in which she waged war was first revealed to a disgusted world," a development with its roots in "the German attitude toward women."[21] This dubious assessment of German and French social structure is calm, however, compared with Christabel Pankhurst's outburst in *The Suffragette*: "This great war . . . is Nature's vengeance— is God's vengeance upon the people who held women in subjection, and by doing that have destroyed the perfect, human balance."[22] It is unclear whether Pankhurst was here celebrating the bloodletting on the western front which was wiping out young men from both sides in truly glorious numbers, or whether she had the Hun specifically in mind. What is clear is that women were as susceptible as men to wartime rhetoric and "holy war" constructions.

Just-war restrictions having been absorbed within the codes of military conduct of European nations, soldiers frequently restrained themselves—even when their nation-state tilted toward unlimited war—under "fair fighting" (*jus in bello*) rules. Indeed, fighting men were often more restrained in actual combat than the abstract rhetoric of civilian war leaders (all male) and propagandists (many of them female) demanded they be. That, however, is a story for the next chapter. For the moment I simply want to register twentieth-century reiterations of the warrior as avatar of justice. Whether volunteering enthusiastically (the hundreds of thousands of British volunteers in the early days of the First World War) or being drafted reluctantly (America's Second World War G.I. Joe, basically a civilian who, if pressed, could rise to the occasion), the vast majority in each warring nation, by all accounts, thought of their soldiers as decent, their cause as just. Without deep and widespread popular support, active complicity from all quarters, representations of the Just Warrior could not crystallize into a sustainable cultural form.

The Attempt to Disarm Civic Virtue

PACIFISM

Those who withheld their support from these grand and sometimes terrible developments in modern Western history—for example, Pacifist sects and "peace churches," Anabaptists, Quakers, the Brethren—never garnered sufficient support to put real pressure on "the system." The historic peace churches have ongoingly witnessed to their conviction that the New Testament offers a "law of love" unambiguously at odds with the old "law of force." But strict, doctrinal pacifism, by which I mean the beliefs and self-definitions of those who prejudge as immoral all wars of any sort for any purpose, has been and remains the purview of a tiny minority in all Western societies. Christian squeamishness concerning militarism, state violence, and the arms race seems to be growing but does not preclude a resort to arms as a tragic necessity: hence, the possible need for Just Warriors.*

Alongside enduring religious pacifism, other pacifisms took root in modern Europe. Less a matter of faith and witness than an article of reason, a variety of free traders, positivist evolutionists, and buoyant rationalists confidently looked forward to a world without war. Pretty badly battered by historic events, these forms of theoretical pacifism are defunct in their more robust forms.† Pacifism historically has drawn women to its ranks, including pre–First World War feminist internationalists and current feminist anti-militarists. Given representations of women as the pacific Other, pacifist identity is perhaps more accessible to women than to men. Pacifist constructions reinforce and reaffirm dominant cultural images of women—Gandhi thought women the best *satyagrahis*—but challenge masculine representations, calling into question male identity as fighters, warriors, protectors.

These suggestions, at least, make intuitive sense and might help explain why women are disproportionately *represented,* in several

* Why the "peaceable kingdom" hasn't lured more into its vision and way of being is a thorny matter, and I take a stab at reflecting on it in chapter 7. The Quakers did have a try at earthly governance in William Penn's "Holy Experiment" in Pennsylvania, where Quaker control lasted seventy-four years (1682–1756). But well before the end, the high hopes that inaugurated the experiment, the certainty that a peaceable kingdom would be established where "Friends could dwell in amity with each other and with the rest of the world," had disintegrated.[23]

† Although, as I observed in chapter 2, the melody lingers on in the form of revolutionary faith that class wars will finally end war—but this outcome will first require a lot of bloodletting.

139

senses, in pacifist movements and as followers of charismatic male pacifist leaders. These women, too, are greatly outnumbered by the majority of their gender who do not enter into pacifist construals as a chosen identification; indeed, women in overwhelming numbers have supported their state's wars in the modern West. The historic representation that invigorates women's collective self-recognition as essentially caring, concerned, nonviolent beings who can, however, be mobilized as wartime civic cheerleaders and home-front helpmeets— "Women of Britain Say—GO!" was the most famous and effective First World War recruitment poster—is the Just Warrior's better half, the Beautiful Soul.*

Female Privatization: The Beautiful Soul

Beautiful Souls are too good for the world yet absolutely necessary to it. Like Aristotle's female householders, they are a necessary condition for, though not an integral part of, the world of free citizens. The narrative of the Beautiful Soul goes like this: insulated through dense historic and social repetitions from the bruising realities of the "actual world" (Hegel's locution), the Beautiful Soul serves as a repository of innocent convictions and self-definitions. Although individual men and women can be Beautiful Souls as discrete individuals—Hegel does not have women as a constituted group in mind when he unfolds the historic consciousness of beautiful souls—women in the West *have* been cast as a collective Beautiful Soul.

---

* The locutions I deploy here, which somewhat mock what mobilized women were and did, are intended to defamiliarize collective identities, removing them from their usual surround and relocating them through redescription. This is a strategy not unlike that of the disillusioned poets of the First World War who mocked received notions of honor and heroism: "What passing bells for those who die as cattle?" asked Wilfred Owen.[24] To call women's flat-out efforts on the home front to sustain the war effort "civic cheerleading" locates these activities as another way for women to exhort and shore up men from the sidelines, no longer with high-school war cries and pompom waving but with grown-up locutions and actions. This takes nothing away from the importance of such experiences *to* women themselves, any more than does Owen's poem from the transformative dislocations of soldiering and battle to soldiers themselves.

The coming together of Beautiful Souls (as a form of group identity) frequently yields "rejoicing over . . . mutual purity, mutual assurance of their conscientiousness" together with expressions of yearnings ardently cast but insufficiently realized to take this collective "her" beyond the need to "preserve the purity" of her heart.[25]* This construction got solidified, as ambiguities and alternatives fell away, by the late eighteenth century when, as Natalie Zemon Davis points out, "absolute distinctions between men and women in regard to violence" had come to prevail.[26] I explore the ongoing entanglement of contrasting feminisms with this cluster of culturally sanctioned identities.

The feminization of particular areas of social life and concern both constrained women *and* empowered them. Women not only played a part in this outcome, but they were, and are, required to affirm it ongoingly (if unreflectively) as social definition gets fleshed out in self-understanding. Embodying ethical aspirations but denying women a place in the corridors of power; recapitulating aesthetic visions of the "lady" unbesmirched by the sordid wheelings and dealings of commercial society, but insulating her from the nameless perils of uncharted social waters, by lodging women solidly in the domain of *Privatrecht*, or "private right," a sphere that persists in tension with the *Kriegstaat*, or "war state"—the Beautiful Soul constellation of enshrined ideas dooms women to lose certain battles over and over again. Pellucid neither to herself nor to others, the Beautiful Soul is custodian and conserver of elemental chords of human memory whose echoes are foreordained to be drowned out by the guns of war, she and others to be temporarily or permanently blinded by the state as an "Apollonian *Lichtgott*, 'the Light-god' who must find fulfilment and self-renewal in war."[27]†

Certain social divisions got sealed as historical preliminaries to bourgeois beautiful souldom: between home life and public life; peace and war; family and state; the immediacy of desire ("the law of the Family is an implicit inner essence which is not exposed to the daylight of consciousness but remains an inner feeling") and the self-conscious power of universal life ("ethical life that is conscious of itself and

---

* This is a bit much, but it speaks to *tendencies*, to *temptations* inherent in the Beautiful Soul's sense of herself. The Just Warrior is subject to similar deformations.

† George Steiner is here discussing Hegel's treatment of the Antigone myth.

actual").[28] Too tidy by far, these formulations reflect and invoke bourgeois society's self-definitions and vehement urgencies, an unstable combination of universalist pretensions and privatizing keenings.

That "real men and women" have never been Beautiful Souls and Just Warriors turned out to meet the abstract specifications of Hegel's systematic philosophy is beside the point. Hegel abstracted from social and cultural forces at work, honing them into dense grandiosities. What he captured was an epoch—the late eighteenth and early nineteenth centuries—in which woman as Beautiful Soul became a constitutive myth explaining, justifying, perhaps even serving as consolation for, women's retreat from sites she once routinely occupied.

Whether as agricultural laborer, tradeswoman, estate manager, sometimes working alongside her husband or her children, sometimes alone or with other women, a medieval woman, for example, held a "full share in . . . private rights and duties" and lived a life that "gave her a great deal of scope, since . . . the home of this period was a very wide sphere."[29] Medieval men and women inhabited a structured but loose-fitting *saeculum* in which distinctions between war and peace, reason and emotion, nature and culture, science and faith, domestic and civil, proper and uncouth, even male and female were, to many intents and purposes, blurred. Involved in nearly all the trades guilds, women worked, marketed, hunted and tended animals, and went on pilgrimage. As there was no separate civic or public sphere as a distinct social form, hence no *citizen* in the modern sense, sharp cleavages between civic and private persons had not yet appeared.

These categories congealed over time with the rise of nation-states, markets and capitalism, the construction of peoples as mobilizable populations.* I shall look briefly at just one important discursive influence: Luther's masculinization of theology and privatization of the self, male and female.

Women are not, in Luther's thought, a special "target of opportunity" for privatization. Instead, the overall effect of his political theology is to strip individuals of involvement in the world of "externals," ines-

---

* I am not here suggesting that we would be well advised to medievalize ourselves—an impossibility in any case. Rather, I am challenging those fond myths that make it possible at this late date for people to believe in History and Progress, grand teleologies deeply entangled in ongoing legitimations of collective violence.

sential—in his view—to the faith of the Christian. What is vital is inner freedom, a continually scrutinized *inner life*. The Christian (Protestant) family as the only secure locus of earthly human hopes was valorized as an arena of compassion, concern, and kindness. But it was also imbued with political definition as the place children learn obedience to authority. Given Luther's strict injunctions to obey secular rule— such obedience cannot compromise the inner freedom of the Christian—his stress on parental authority is an urgent necessity. If the child does not learn obedience in the home, the temporal order will be threatened. Not yet the Puritan's "little commonwealth," Luther's family is unmistakably laced through with politicized imperatives.

Perhaps even more significant in the long run, to a consideration of culturally sanctified semiotics of womanhood, is Luther's deconstruction of potent and authoritative female images. He assaulted the Madonna repeatedly, blasting exaltation of Mary as "one of the chief sins of the Romanists."[30] Female saints (and their male colleagues) were collective casualties of this cultural and political war on the medieval tradition, as was the feminized construction of *mater ecclesiae* herself. Promulgating a theology from which the dominant female symbol had been expunged, Luther set in motion, and was himself enmeshed in, that concatenation of forces that gave rise to secular male dominance. His masculinization of theology played an important role as the opening salvo in securing the political theology of the Protestant nation-state.

With veneration of *the* succoring and interceding Mother (rarely thought of as a *wife*) trivialized as "ignorant idolatry," "the female" as a charged repository of *human* hope faded. The medieval Madonna had *not* served primarily as a sign of female privatization; rather, she sanctioned compassion, promised surcease, and dignified earthly sorrow, a routine expectation for human beings living on the edge of mortality in centuries visited by flood, fire, famine, Vikings, and the plague. Neither a sentimental nor a sexless figuration, the Madonna was often portrayed with a breast exposed, being suckled or nuzzled by the Babe. This expression of *maternal* embodiment also retreated into the background, to be superseded by Enlightenment-inspired images of the female (nature) exposed to, and by, the scrutinizing eye of the male (culture). No longer anchored to a semiotics of female power

143

and transcendence, increasingly located *out* of the world, "the woman" is deployed as a sign of enshrined conservatism, her consciousness at once entombed *and* incited by bourgeois valorizations—as a brief examination of the ways women marked their identities in and through Beautiful Soul constructions helps make clear.*

Western women have devised various means to *realize* what they in some sense were through cultural definition. Some have sought to bring the ethics of which they were cultural guardians to bear on the bourgeois social and political world. Pictured as frugal, self-sacrificing, at times delicate, the female Beautiful Soul in time of war has been positioned as a mourner, an occasion for war, and a keeper of the flame of nonwarlike values—and has thus been set up as a being, and a whole way of life, men both cherish and seek to flee, both need and despise. For nonwarlike values can take generous forms—sexual love, family devotion, play and contemplation, the aesthetic articulation of a way of life—and constricting ones—moralistic restrictions on simple pleasures, hygienic crusades, rule-bound behavior, insularity of outlook. In peacetime, and through civic action, women may try to make the "outer world" of the *civitas* resemble an image of the perfect home— ordered, healthful, clean, comfortable.†

My argument is *not* that women have set as an explicit, self-conscious task the holding intact of cultural repetitions with which they have been unavoidably entangled. Rather, more subtly, I find fascinating the ways prior constructions have been shored up even when the *explicit* aim of a text, or broadside, or speech, or slogan was to challenge received understandings of woman and her place. During the struggle over women's suffrage, as I indicated in the introduction, suffragists and their opponents often *shared* the same metaphoric and metonymic turf, repeating locutions, reinscribing symbols, evoking tropes that locate men and women in ways that circumscribe rather than enlarge their range of possibility.

Wartime presents many perplexities, for women often engage in tasks recently denied them as they enter occupations previously closed

---

* Overstated and one-sided, this is a side that needs to be brought out and pinned down in light of the fact that new sightings of the Beautiful Soul can be made even in our very different social world.

† In the United States, several generations of women activists burst out of the house and into urban progressivist reform efforts—a domesticizing strategy that goes back as far as the Greek comedies of "reversal."

to them and take risks from which they have been protected: for example, by becoming teachers (the Civil War gave American women their first real opportunity to teach in large numbers), nurses in dangerous places, industrial laborers, sexual rebels. These social realities, however, need not swamp reigning symbolic constructions that are constitutive features of women's identities.

If one turns to female narratives, one way to expose their repetitions is by interrogating texts, investigating whether Beautiful Soul constructions are being shored up or displaced.* Pertinent questions for such an interrogation would include: (1) Does the author define all women in opposition to all men? (2) Do the author's rhetorical choices invite self-congratulatory responses and lend themselves to sentimentalist reactions? (3) Does the author open or foreclose space for debate and disagreement? Is it possible to challenge her assumptions or conclusions in good faith, or is one required to adopt the voice of an apostate? (4) Does the author compel us to think in the absence of certainty or ensure certainty at the cost of critical reflection? (5) Do the author's formulations *reassure, soothe, bring relief to, reinforce, reaffirm*; or do they *disturb, unsettle, take apart, make ambiguous*? (6) Is the author's voice didactic, ironic, moralistic?

My candidates for this interrogative exercise—to be conducted by the reader, hence presented without interpretation by the writer of this narrative—are texts of various sorts by American and British women, most pirouetting around war:

1916. From a banner hoisted by suffragists on parade before the Republican National Convention

> For the Safety of the Nation To the Women Give the Vote
> For the Hand that Rocks the Cradle Will Never Rock the Boat!

1916. Countess of Warwick, *A Woman and the War*

Of late years a certain number of women of all classes have been drinking more than is good for them, and since the war broke out the working women's temptations in this direction and the opportunity to indulge them

---

* One does not ask the author's *real* intentions—a concern irrelevant to the task of determining whether the discursive trail markers erected by a particular author take one back, repeatedly, to the "same" place, especially if she is self-defined as a social critic.

have grown side by side. . . . The majority of working women are as sober as the majority of every class, but . . . they are matched by thousands of intemperate ones . . . and I feel they should be saved from themselves. . . . The intemperate classes may resent restrictions, but it remains necessary in their own interests. . . . It is her [woman's] mission in this world to sacrifice herself, from the hour when she accepts motherhood until the end. . . . She has found the uses of adversity, she has accepted self-sacrifice for the sake of those who will be the better able to enjoy the rich fruits of life. . . . Motherhood enforces the qualifications of women, justifies their claims and provides them with the material to train for future triumphs. . . . At no period in the history of Western civilisation, has it been more necessary for the women who count as factors in world progress to consider their duty and fulfil it to the extreme limit of their power. . . . Black and yellow races alike are extraordinarily prolific; there is among their women no shirking of duty in that regard.[31]

1918. "The Ten Commandments of Womanhood," "Prepared by the President of the Connecticut Congress of Mothers" (see facing page).

1918. Mrs. F. S. Hallowes, *Mothers of Men and Militarism*

Women equally with men have a passionate love of mother-country. . . . Though we loathe slaughter we find that after men have done their best to kill and wound, women are ever ready to mend the broken bodies, soothe the dying, and weep over nameless graves! . . . God made two, a man and a woman, to rule the home—the state is but a larger home. . . . A mother's love is at its best the most *selfless* love of humanity. . . . If all the school histories were collected and cremated, and replaced by a series of international text books compiled under the direction of an international committee, a great forward step would be taken in the direction of decency and true education.[32]

1982. Barbara Zanotti, "Patriarchy: A State of War"

Why weren't we prepared for this?—the imminence of nuclear holocaust; the final silencing of life; the brutal extinction of the planet. . . . We have lived with violence so long. We have lived under the rule of the fathers so long. Violence and patriarchy: mirror images. An ethic of destruction as normative. Diminished love of life, a numbing to real events as the final consequence. We are not even prepared. . . . Wars are nothing short of rituals of organized killing presided over by men deemed "the best." The

The Ten Commandments of Womanhood

NINETEEN HUNDRED AND EIGHTEEN

Thou Shalt Not Waste Time, for idleness is shame and sloth a mockery; and lo! the day cometh when thy men shall be called from the harvest and their workshops stand empty and silent.

Thou Shalt Not Waste Substance, for once, thrice and ten times shall thy country call upon thine household for gold, and woe betide the land if at the last thy purses be found bare.

Thou Shalt Not Waste Bread, for every fragment that falls idly from thy board is withheld from the mouths of thine allies' children, and the kits of thy sons and brothers in the trenches.

Thou Shalt Not Bedeck Thyself Lavishly, for the silk upon thy back and the jewel upon thy breast are symbols of dishonor in the hour of Earth's agony and thy nation's peril.

Thou Shalt Not Be Vain and Self-Seeking, for the froward and jealous heart judgeth itself in the sight of the Lord; and in the time of world travail who shall say to her sister, "I did it and thou didst it not."

Hearten Thy Men and Weep Not, for a strong woman begetteth a strong man, and the blasts of adversity blow hard and swift across the world.

Bind Up the Wounds of Thy Men and Soften Their Pain, for thy presence by the light of their campfires is sweet and grateful, and the touch of thy hand deft in the hour of need.

Keep Thou the Faith of Thy Mothers, for in the years of thy country's sacrifice for Independence and Union they served valiantly and quailed not.

Keep Thou the Family Fruitful and Holy, for upon it the Lord shall rebuild His broken peoples.

Serve Thou the Lord Thy God with Diligence, for His houses of worship shall not be empty nor His altars unvisited, in the years of His mighty chastening.

Prepared by the President of the Connecticut Congress of Mothers
Issued by the Connecticut State Council of Defense

fact is—they are. They have absorbed in the most complete way the violent character of their own ethos. . . . Women know and feel the lies that maintain nuclear technology because we have been lied to before. We are the victims of patriotical lies. . . . To end the state of war, to halt the momentum toward death, passion for life must flourish. Women are the bearers of lifeloving energy. Ours is the task of deepening that passion for life and separating from all that threatens life, all that diminishes life; becoming who we are as women.[33]

1985. Barbara L. Baer, "Apart to the End?"

Male writers view technology and weapons in a way women do not. . . . Men have had a relationship with weapons over the centuries that women have not. . . . The history and technology of its weaponry still fascinate him. I say this not to condemn the male writer but to show how he takes part of a masculine point of view. Women, on the other hand, tend to ignore weapons and even leave out description of war itself. . . . And while men write about the effects of war, they intellectualize them in a way women do not.[34]*

These excerpts, though selected for the purpose of interrogation, are not unrepresentative. Examples abound, from suffrage leader Carrie Chapman Catt's declaration on the eve of the Second World War that women are "devoid of the war spirit" and men "have made all the wars in history,"† to the militant Alice Paul's insistence, in 1941, that war springs from man's nature, for "women's instincts are constructive. . . . [The guilt of war can be] laid wholly on men."[36]

Beautiful Souls and Just Warriors, though knocked about, continue (in and through rhetoric brought up to date) as ready-made identities which become particularly compelling, or are made possible only for men, in time of war. And many human beings yearn, secretly and silently, perhaps unaware, for a "war." I mean, that is, they dream dreams of a period of individual and collective testing and *esprit*. In the throes of war, the individual is taken over by a *force majeure*, drawn

---

* Unable to resist altogether the siren call, I offer the suggestion that Baer alone locates her discussion in a way that, although it makes contact with, it does not get stuck within, the Beautiful Soul labyrinth.

† Mary Beard commented acerbically, with reference to Catt: "In eight words" she rhetorically cleared "women of all war guilt. . . . With that innocence it appeared logical that women . . . were inclined to peace by their very make-up."[35]

148

out of the daily grind, absorbed within a wider, swifter stream, does something that "really counts," finds out if he or she has what it takes to fight (or to refuse to fight), to sacrifice, to endure, to triumph, or to survive defeat.

That this *isn't* what it would *really* be like scarcely matters: it would, or has been, enough like that to keep the dream alive, and stubbornly resists dissolution in light of nuclear realities, "home front" vulnerabilities, Vietnam dislocations. "History teaches" us (the collective female "we") that Beautiful Souls have not only not succeeded in stopping the wounding and slaughter of sons, brothers, husbands, fathers, but have more often exhorted men to the task, sustained their efforts, honored their deeds, mourned their loss. But history does not teach; rather, we "teach" *it* by making it "speak" to us in various ways, by remembering this and forgetting that. One form this attempt to teach history currently takes is revitalized just-war thinking.

Implications of the Just-War Tradition

> Christians, even as they strive to resist and prevent every form of warfare, have no hesitation in recalling that, in the name of an elementary requirement of justice, peoples have a right and even a duty to protect their existence and freedom by proportionate means against an unjust aggressor.
> —*Gaudium et Spes* (the Pastoral Constitution on
> the Church in the Modern World, a Vatican II document)

> Given the view of Nazism that I am assuming, the issue takes this form: should I wager this determinate crime (the killing of innocent people) against that immeasurable evil (a Nazi triumph)?
> —MICHAEL WALZER, *Just and Unjust Wars*

One way to tell the story of just-war discourse is to treat it as an authoritative tradition dotted with its own sacred texts, offering a canonical alternative to realism as received truth. Rather than beginning with Machiavelli (or, reaching further back, Thucydides), just war as continuous narrative starts with Augustine; takes up a smattering of medieval canonists; plunges into the sixteenth century with Luther as the key figure; draws in a few natural/international–law thinkers (Francisco Suarez, Francisco de Vitoria, Hugo Grotius), then leapfrogs into the era of modern nation-states—and wars. At that historic juncture, with religious and philosophic discourse severed, just war becomes a partial preserve of theologians. Catholic thinkers predominate, but Protestants also figure. The professionalization of moral and ethical discourse, similar in this sense to that of international relations, spawns essays and monographs on the moral philosophies of deontology versus consequentialism as applied, for example, to such questions as the rule of proportionality and weighing problems of collateral damage, which are primarily accessible to other professionals: this has been a partial fate of just-war thinking.*

Canalized as just-war *doctrine* historically and *theory* currently, just-war teaching is sometimes presented as a cluster of "Thou shalts" and "Thou shalt nots," Augustine's rich ruminations giving way to a list of seven (or more or less) requirements presumed operative on states, statesmen and women and war fighters. These are: (1) that a war be the last resort to be used only after all other means have been exhausted; (2) that a war be clearly an act of redress of rights actually violated or defense against unjust demands backed by the threat of force; (3) that war be openly and legally declared by properly constituted governments; (4) that there be a reasonable prospect for victory; (5) that the means be proportionate to the ends; (6) that a war be waged in such a way as to distinguish between combatants and noncombatants; (7) that the victorious nation not require the utter humiliation of the vanquished.[37]

Just-war thinkers as layers-down-of-the-law insist that they are not propounding immutable rules so much as clarifying the circumstances that should—and actually, if imperfectly, do—justify a state in going to war (*jus ad bellum*), and what is and is not allowable in fighting the wars to which a state has committed itself (*jus in bello*). Cast in this form, modern just-war discourse downplays some features of the human condition and the Western tradition and brings others into sharp and prominent focus.

---

* There isn't much of a female presence on the professional side of these developments for reasons that need no explaining. Female voices were *heard* by the Catholic bishops as they conjured with their 1983 pastoral "The Challenge of Peace" but were not, by definition, among its *authors*.

The Attempt to Disarm Civic Virtue

With their aims of constraining collective violence, chastening *realpolitik*, and forging human identities, the current heirs of this way of thinking assume (1) the existence of universal moral dispositions, if not convictions—hence, the possibility of a nonrelativist ethic; (2) the need for moral judgments of who/what is aggressor/victim, just/unjust, acceptable/unacceptable, and so on; (3) the potential efficacy of moral appeals and arguments to stay the hand of force. This adds up to a vision of civic virtue, not in the classical armed sense but in a way that is equally if differently demanding.

One brief example: the American Catholic bishops, in common with all Christian just-war thinkers, do not locate survival as an absolute, hence do not accept retaliation as an acceptable way of "fighting back." It follows that the use of nuclear weapons for the purpose of destroying population centers is to be condemned *under all circumstances*, even if one's own cities have been the target for a first strike. The capacity to *refrain* from violence when one has been the victim of it is central to pacifism, of course, but it also lies at the heart of just-war thinking which shares with pacifism the insistence that violence, not nonviolence, must bear the burden of proof. Although combat soldiers learn the harsh lesson of *restraint* in the vagaries of actual fighting, noncombatants do not experience the need to refrain in the same way.

One can imagine two scenarios: in the first, survivors of a first-wave assault on key American cities, thirsting for revenge, demand that the perpetrators be "punished"—indeed, "wiped out."* In the second, survivors, having been apprised in the most horrendous and visible way of the horrors of nuclear war, call for forbearance, not retaliation: they become peace witnesses like the survivors of Hiroshima or Nagasaki. That there is historic warrant for this alternative scenario comes from two recognitions, one abstract, the other concrete: first, that just-war thinking as popular morality cuts across the West, offering the possibility for sober reflection by ordinary men and women on questions of violence; and, second, that opinion polling in Great Britain in 1941 showed that "the most determined demand for [reprisal raids

---

* This scenario is probably fanciful—not because a lust to annihilate one's tormentors would not arise, but because it seems unlikely that there would exist the mechanisms for, or the possibility of, citizen reaction feeding into decisions by statesmen and women in the disordered aftermath of a nuclear attack. By the time citizens had gotten over the shock and registered their vehemence, the deed would no doubt have been done.

against Germany] came from . . . rural areas barely touched by bombing, where some three-quarters of the population wanted them. In central London, conversely, the proportion was only 45 percent." Michael Walzer concludes that "men and women who had experienced terror bombing were less likely to support Churchill's policy than those who had not—a heartening statistic."[38] I agree.

Just-war thinking as a form of civic virtue cannot endorse the unleashing of aggressivity sanctioned by armed civic virtue in a time of total war. Indeed, what is demanded instead is *deep reflection* by Everyman and Everywoman on what his or her government is up to. This, in turn, *presupposes* a "self" of a certain kind, one attuned to moral reasoning and capable of it; one strong enough to resist the lure of seductive, violent enthusiasms; one bounded by and laced through with a sense of responsibility and accountability. In other words, a *morally formed civic character* is a precondition for just-war thinking as a civic virtue. Determining whether and to what extent this possibility *really* exists is not my task at the moment; rather, I shall look briefly at current ferment from persons religious and theoretical working within a just-war frame.

CATHOLIC CONTROVERSIES:
THE BISHOPS' 1983 PASTORAL LETTER

On one end, the Catholic just-war tradition makes contact with pacifism; on the other, it elides into holy wars against infidels. If medieval Catholicism gave us the Crusades, contemporary Catholicism edges nearer the pacifist pole, sanctioning pacifism as personal identity and collective witness without embracing it wholesale, and rejecting unequivocally crusading enthusiasms. All modern popes—from Pius XII who condemned total war *during* total war, repudiating the terror bombing of German and Japanese cities with the means of conventional war; to John Paul II, who has been outspoken on nuclear arms, the arms race, and the distortion of science given its complicity in building up "arsenals of destruction"[39]—have situated themselves as spokesmen for peace, as nuclear but not absolute pacifists. Thus, they find no moral justification any time anywhere for the *use* of nuclear weapons, no matter what the alleged provocation, danger to oneself, or wounding by another.

War is still "permissible" if all peaceful alternatives have been exhausted in serious efforts; if the war is waged under *jus in bello* rules which prohibit making war on noncombatants, hence rule out much of the panoply of modern war; and if war is a response to aggression or the only way to protect the innocent from "certain" harm. These statements add up to a powerful construction proffered as an occasion for reflection *and* as prohibitive moral rules: "Thou shalt not. . . ." Papal pronouncements also affirm Gospel injunctions against punishing one's foes once you have beat them: there shall be no postwar retaliation nor vengeance against the defeated. After war, those who have prevailed must show the face of mercy and forgiveness rather than bare the talons of retribution.*

Nor are Catholic bishops latecomers to just-war reflection and teaching. The 1983 pastoral letter issued by the American bishops is the latest in a long train, reiterating the nuclear pacifist position, condemning total war, instructing statesmen and citizens on war and peace within a just-war frame. The strategic debate of recent years has focused in part on deterrence. The bishops insist that the United States's current strategic doctrine is unacceptable on moral grounds, as it targets civilian populations; and, as well, that deterrence is structured in such a way that arms buildups are its automatic outgrowth given the "worst case scenarios" that guide and feed it. The proposition *Si vis pacem, para bellum* ("To have peace, you must prepare for war"), always debatable, now lacks any credibility (we have seen and know too much), functioning instead as a whitewash of one's own moves to gain the upper hand.

The enormous flap created by the bishops' pastoral letter arose in part from the widespread insistence that it is not the bishops' job to pronounce on such hefty matters as nuclear strategy—but is, instead, a task for the experts. The bishops listened to hours of expert testimony, only to be blasted, by some critics, for not being themselves strategic-deterrence professionals. By what right, then, did they speak? The need to shore up the epistemological privilege of professional elites

---

* Located at the heart of the Gospel message, forgiveness is mostly forgotten in the political affairs of men—and women. Hannah Arendt calls forgiveness Jesus of Nazareth's most radical and important teaching—"the exact opposite of vengeance," which sets in motion chain reactions. Jesus' teaching contains a message of freedom, "freedom from vengeance, which encloses both doer and sufferer in the relentless automatism of the action process."[40]

comes through loud and clear in such criticisms, showing just how defensive the defensive establishment has become. Telling the bishops to shut up and stay inside their churches signifies the continued simplistic insistence that religion and politics must not mix.

Politics as policy formulation and implementation is not for amateurs. Women, too, are well advised to keep their noses out of this complex business *unless* they have learned not to think and speak "like women"—that is, like human beings picturing decimated homes and mangled bodies when strategies for nuclear or other war fighting are discussed. The worlds of "victims"—overwhelmingly one of women and children—and of "warriors," as Freeman Dyson characterizes them,[41] have become nearly incommensurable universes to one another. Contemporary rationalist realism freezes this division. Although images of male fighters and female noncombatants have been the usual expectation, just-war discourse at present tends to see both soldiers and civilians as likely *victims* of future wars and places men and women in all walks of life under the purview of its teaching as practical morality. Thus the pastoral letter addresses: Educators, Parents, Youth, Men and Women in Military Service, Defense Industry, Science, Media, Public Officials as well as Citizens.

JUST WAR AS POLITICAL PHILOSOPHY: MICHAEL WALZER

Michael Walzer's *Just and Unjust Wars* (1977), a reformulation and justification of just-war thinking that shares some but not all of the ontological and moral commitments just discussed, consists of sets of presumptions and arguments that require consequentialist assessments. Walzer rejects pacifism outright, being far less open to its message than the bishops, for example, and leaves a crack that opens onto the door to crusading enthusiasms that theologians have tried to slam shut. Walzer goes this route because he locates the survival and freedom of political communities—states—as the highest value. For Catholic thinkers, contrastingly, the rights of states to an ongoing existence is a pre-eminent but not a penultimate good. As a "book of practical morality," in his words, Walzer's text offers complex cases involving often fine distinctions.[42] The insistence that violence is always regrettable and devoutly to be avoided holds. But many horrors slip through for Walzer that do not for other just-war thinkers: for example, his treatment of British decisions to bomb German cities during the Second World War, and nuclear deterrence theory, revolve around means-ends calculations in tension with classic just-war prohibitions on targeting civilians.

Thus he justifies the saturation bombing of German cities—Dresden excepted—given the *nature of the Nazi* threat and the predictable outcome should Britain fall to Germany. Present threat and future danger fuse to override *jus in bello* rules. The British made total war on the most densely populated areas of major German cities, some of them not "military" targets in any compelling meaning of the term. By constructing Nazism as an *immeasurable* evil, not just *an* evil, Walzer must put his question concerning terror bombing this way: Should I wager a determinate crime against an immeasurable evil? A condition of "supreme emergency"—the Second World War—erodes the force of just-war restrictions in his argument. By adjusting to the exigencies of total war, Walzer slides into conclusions a "modified realist" would probably be able to live with. Concluding that our present circumstance is one of continuing "supreme emergency," Walzer, in a schematic discussion, finds nuclear deterrence criminal but "unavoidable": not a very good way "to cope" but cope we must "in a world of sovereign and suspicious states."[43]

Because Walzer owes so much to modern constructions of "rights" engrafted from one tradition—liberal, rationalist, litigious—his language is sometimes inadequate to his task. Opposing American use of atomic bombs on Hiroshima and Nagasaki, for example—finding them unjustifiable within his frame—Walzer goes on to ask: "How did the people of Hiroshima forfeit their rights?"[44] The language of rights is impoverished in this context, inadequate to capture what happened on those dreadful days. Pope Paul VI got closer to the mark when he spoke of a "butchery of untold magnitude," and drew upon prophetic language, which Walzer avoids just as he eschews vivid descriptions of wartime carnage, making no attempt to evoke actual war experience. Walzer's "practical morality" seems a bit too abstract, too finely tuned, to guide most of us, bleaching out the texture of historic experience as it does.*

---

* The same criticism can be leveled at the bishops' pastoral letter, but the letter can more readily be translated into homey reflections than can Walzer's highly schematized discourse. I

ARMED CIVIC VIRTUE

THE MODERN DILEMMA

The great strength of all this moral stocktaking is the fact that it begins with presumptions that challenge the use of force—hence, all acts of collective and individual violence. Constituting men and women as *concerned citizens*, just-war thinkers enjoin them, whatever their respective vocations, to get serious about matters of war and peace. But how does this "play in Boston"? Brought down to parish and popular levels, the resurgence of just-war thinking offers moral dilemmas and provides clear-cut or provisional answers that are urged upon others.[45]

In actual wartime, it seems likely that many of these stipulations will go unheeded. They *may* have sufficient force to give warmakers a bad conscience, and they offer men and women a "voice" in and through which to register their objections and their moral distress. That just-war teaching *has* made a difference in concrete situations and reactions is clear. *Commonweal*, a Catholic lay journal, condemned terror bombing in 1942, as did other Catholic journals of opinion. Reinhold Niebuhr, cohabiting at times with just-war views, called the total-war mentality "nauseous self-righteousness." The atomic bombings were an occasion for sorrow and criticism on the part of many leading Protestant and Catholic journals of any note. For example: in its 24 August 1945 issue, *Commonweal* concluded its editorial condemning the atomic bombs in these words:

> For our war, for our purposes, to save American lives we have reached the point where we say that anything goes. That is what the Germans said at the beginning of the war. Once we have won our war we say that there must be international law. Undoubtedly. When it is created, Germans, Japanese, and Americans will remember with horror the days of their shame.[46]

On the other hand, the available public-opinion data suggests that, in early polls, 80 to 85 percent of Americans, most "good Christians,"

here tap some of the concerns and criticisms my students have expressed upon reading Walzer's book, and am especially grateful for the thoughts of Ken Bertsch, my graduate teaching assistant.

endorsed use of the atomic bomb,[47]* *pace* the expressed views of leading articulators of the just-war position. According to Paul Boyers, letters to the editor at the time of the bombings frequently expressed regret "that atomic bombs had not been used to destroy all human life in Japan. A Milwaukee woman expressed her genocidal impulses this way: 'When one sets out to destroy vermin, does one try to leave a few alive in the nest? Certainly not!' A letter in the *Washington Post* on August 17 from a woman who said the atomic bomb made her ashamed to be an American elicited a torrent of bitter, abusive letters."[48] Exterminationist rhetoric was rife, and support for a policy of annihilation against the Japanese was widespread. One 1943 "best-seller stated that the fight against Japan had to continue 'until not alone the body but the soul . . . is annihilated, until the land . . . is plowed with salt, its men dead and its women and children divided and lost among other people.' Carthage, sacked and razed by the Romans struck the more historically minded as an apt model for Japan. . . . *Collier's* ran an editorial entitled 'Delenda est Japonia.' "[49]

How to explain this popular savagery? In part, it stems from the crusading impulse lurking in the interstices of just-war discourse. The language of good and evil, just and unjust, may, under conditions that invite total war, turn people out as judges who sometimes become executioners. Should this happen, the preponderant force of "official" just-war discourse may not have sufficient strength to hold in check the "heretical" crusading offshoot. In a pithy critique of the moral absolutism of pacifists, Reinhold Niebuhr argues that the Christian faith "ought to persuade us that political controversies are always conflicts between sinners and not between righteous men and sinners. It ought to mitigate the self-righteousness which is an inevitable concomitant of all human conflict."[50] Hitler, or rather our collective and ongoing reactions to him, continually imperil such recognitions. Hitler gave the Allied countries permission, so to speak, to indulge in collective self-righteousness. Constructed as the *incarnation* of Evil, a kind of mirror-image demiurge of the "good god," Hitler worked, and

---

* At the time, my mother was among the small minority (some 15 percent) who disapproved the dropping of the bomb, and she still thinks it was wrong. My father, no warrior, had mixed feelings at the time, but figured the bomb's use was dictated by military necessity, that thousands of American boys would have died if we had had to attack the Japanese mainland.

works, wonderfully whenever we want to reconfirm our benign sense of ourselves. By 1945, we were morally inured to obliteration of cities, and use of the atomic bomb was an extension of what had become standard war policy.

The force of these reflections is not to suggest that just-war prohibitions must go by the board given war excesses, but that they are under terrible pressure to succumb. Presupposing the possibility of civic virtue (partly) disarmed, just-war teaching presumes that all human beings can and should attempt to thread their way through the violent currents of the past, present, and projected future.

To make more secure contact on the level of "lived life," to enter into the fray from the ground up rather than descending from the lofty pinnacle of a "morality system," just-war thinkers would do well to retrieve Augustine's way of thinking as part of an effort to capture the living textures within which limited human beings think and act. Augustine deflates rather than builds up the possibility that we might one day control events. No human being can foreordain endings, happy or otherwise. We cannot even safely foretell what we shall do "on the morrow." Augustine's awareness of the fragility of life in the *saeculum*— no tidied-up world of foreordained, abstract duties and obligations, but a hothouse of conflicting pressures—his way of *evoking* bitter conflicts between real protagonists, his biting irony, and his lyrical descriptions—all speak to his attunement to human shortcomings and tragedies.

Inviting neither total relativism nor despairing withdrawal, Augustine's tragic recognitions point to modes of moral thinking stripped of the demand for triumphant moral heroism. He seeks to limit the damage done, by oneself and others, rather than to preach an unattainable counsel of perfection that invites smugness and despair. The distance that separates Augustine from the Catholic bishops or Walzer is not merely a question of chronology but a matter of where one *locates* one's hopes and what these hopes require. Walzer presumes and requires that human beings act as moral judges who weigh the consequences of their deeds against the magnitude of others' acts, arriving at judgments of blame or (relative) innocence. The bishops presume and require that human beings be prepared to do the "right thing" even when others have done them wrong.

The Attempt to Disarm Civic Virtue

Is this degree of civic virtue possible in the modern world? Are most human beings in the West open to such constructions and prepared to take them on as their own? Just-war analysts would say human beings in the West have already taken them on—thus justifying their (these analysts') arguments. But suppose the analysts are wrong, at least as the propositions are usually stated. It was from a strongly felt moral perspective that Robert Kennedy—as attorney general, the President's brother, and a member of the team handling the October 1962 confrontation between the United States and the Soviet Union over the latter's moves to install ballistic missiles in Cuba—urged caution during the Cuban missile crisis, specifically arguing against a U.S. air strike on Cuba. This stance was later called by the former secretary of state Dean Acheson, who had been brought in as an adviser attending the meetings, "emotional intuitive responses more than . . . the trained lawyer's analysis." Kennedy did *not* offer his caveats as a just-war perspective or in those terms; rather, he insisted that a "surprise attack by a large nation on a small one, entailing many civilian casualties, would go against America's traditions."[51] Just-war moral principles (noncombatant immunity) seeped into his pronouncements to become resonant and compelling only because, as a "Pearl Harbor" in reverse, the proposed air strike would cut against the *American* grain. Where some members of the crisis team advocated a surprise air attack against Cuban missile sites, Robert Kennedy passed a note to the President which read: "I now know how Tojo felt when he was planning Pearl Harbor."[52]

This example suggests that the structure of a particular nation's history and experience will be more salient to political decision makers who enter into debates of moral principles than will be finely honed ethical systems. Similarly, for ordinary men and women caught up in structures of wartime killing and dying and peacetime childbirth and rearing, the experiences of soldiering and mothering have been occasions both for reflection, for constructing narratives that tell "my story," and moments of silence when their recognitions fall outside the frame of acceptable, articulated meanings.

PART II

LIFE GIVERS/
LIFE TAKERS:
HISTORY'S
GENDER GAP

5

Women: The Ferocious Few/ The Noncombatant Many

At first the war is like a holiday at [college]: everyone runs and shouts and laughs and jumps up and down to think that America is finally in it. . . . Then the boys begin to have a different look, as if the real things are happening somewhere else now, and college isn't so important, not any more. They look at you across a kind of gap: you aren't going where they are going; you'll just be studying. The war makes the boys more important and you can see it. People treat them differently.

         —NANCY PRICE, *An Accomplished Woman*

When we Chinese girls listened to the adults talk-story, we learned that we failed if we grew up to be but wives or slaves. We could be heroines, swordswomen. Even if she had to rage across all China, a swordswoman got even with anybody who hurt her family. Perhaps women were once so dangerous that they had to have their feet bound.

         —MAXINE HONG KINGSTON, *The Woman Warrior*

THIS CHAPTER is not a historical account in the usual chronological sense. Instead, I interweave tales from the past, fragments of remembered or forgotten words, creating a narrative with room for many characters to fit—not always comfortably—inside. My aim is to loosen up the congealments of the present as these get reinforced by a past we cannot avoid as historic beings. The assumptions to which

I refer are those that locate women in an auxiliary and pacific role in relation to war in the modern West, leaving the war fighting and dying to the men. Though women may be war's victims, they are, in this reigning narrative, neither its initiators nor its perpetrators. Abstractions and details, commentaries and quotes proliferate. In musing on life giving and life taking as prototypical female and male activities, respectively, I found many voices clamoring for attention. Some shouted; others whispered; several wept. While some of these voices challenged certain of my own assumptions, I have not tried to shut them up. For there are sanctimonies to deconstruct, amnesias to lift, stories to remember.

The Historic Cleavage

In 1982, the *Women's Studies International Forum* turned over an entire volume to one topic: *Women and Men's Wars*. I wrote a piece for inclusion, an early rumination on Beautiful Souls and Just Warriors.[1] But I have puzzled, from time to time since, about the naming of the problem in a way that signified to the reader not just *what* he or she was about to read but *how* to read. The title directs us toward a reinscription of shared cultural expectation: wars are men's to which women are conjoined by the humble, overused particle *and*—a term that connects but simultaneously apostrophizes. An apostrophe for us is a mark that signifies an omission. In Greek usage, an apostrophe was a *turning away*. In the matter of women and war we are invited to turn away. War is men's: men are the historic authors of organized violence. Yes, women have been drawn in—and they have been required to observe, suffer, cope, mourn, honor, adore, witness, work. But the men have done the *describing* and *defining* of war, and the women are "affected" by it: they "mostly react." A benediction is effected rather than a valediction pronounced: *It has always been thus*, we are invited to remember. It will continue to be thus—unless we (women) stop it.

Women: The Ferocious Few/The Noncombatant Many

On some level this formulation rings true. It is women and men's wars. But this is a partial truth flowing from the fact that we have compelled the remembered past to run in a few channels headed in directions already known to us, so that we won't be surprised at outcomes, even if we don't like them very much. Thus: To men's wars, women are backdrop. To women's homes and babies, men play similarly supporting roles: a man's involvement in paternity is inferential given his role in the procreative process. He may own the house, but she makes the home—if anybody does. Women's involvement in war seems to us similarly inferential, located somewhere offstage if war is playing.*

For the most part, we accept some rough and ready division between male life takers and women life givers, a cleavage enshrined through such symbolic vehicles as Beautiful Soul and Just Warrior. We tend to think that the scission expressed in "men's wars, women and" has always been and will always be. Maybe savages (uncivilized) tribes or peoples, now or in the past, behaved differently—but they do not count for much and certainly offer no alternative to the dominant narrative.

Viewing themselves through the lens of this construction, men see edifying tales of courage, duty, honor, glory as they engage in acts of protection and defense and daring: heroic deed doing. Women see edifying stories of nobility, sacrifice, duty, quiet immortality as they engage in defensive acts of protection, the nonheroics of taking-care-of. To be sure, total war has thrown irritants into the refinements of this picture, and the terrible visions of nuclear war with which we have become sated turn us all out as doomed victims. But images of leveled cities, refugees clogging highways, starvation, and disease are snapshots from another place for Americans, foreign, unfamiliar: it can't/hasn't happened here.

Male and female bodies figure centrally in our shared reiterations of women and life giving, men and life taking. Male bodies, straight, hard, are more fit for combat and toughening. Male bodies are also

---

* A difference, of course, is that males are co-initiators of the procreative process. It is astonishing how little this seems to matter when women get together to talk about pregnancy and childbirth as a mysteriously self-inclosing *force majeure* to which men, even the involved ones, are exterior. More on this and the self-enclosing mysteries of war later.

expendable in large numbers, from the First World War's "fallen" to Vietnam's "wasted." The dominant sex is also the one Western cultures have most enthusiastically bumped off. This seems odd at first blush. But consider female bodies—softening, rounding out, *giving* birth. The bodies of young females are not expendable: they are what re-creates and holds forth promise of a future. The body of the young woman is not yet complete. She needs time to *give* birth. Then the men may take that life away if the life brought forth is of one like themselves. (The women get "taken," too, but not to be destroyed.)* Popular understandings of female givers, male takers, loft upward, becoming narrative truth for many, including contemporary scholars, male and female, and (some) feminists. Women and *men's* wars; we are reassured. Whatever women may finally do once wars have begun, women don't start them.† Men are the first cause, the prime movers, of war. The finger of rhetorical blame zeroes in on its target.

When I pointed that finger, I found it turning back toward me— suggesting a strategy that does not assume "men's wars, women and." For wars are not men's property. Rather, wars destroy and bring into being men *and* women as particular *identities* by canalizing energy and giving permission to narrate. Societies are, in some sense, the sum total of their "war stories": one can't think, for example, of the American story *without* the Civil War, for that war structured identities that are continually reinscribed. Whatever else it may be, war is *interesting* (*inter-est* being, as Hannah Arendt reminds us, that which lies between people, relating and binding them together).[2] The soldier has been to hell and back, and that fascinates. One of the endlessly rekindled vexations of anti-war activists and peace philosophers is the fact that peace does not enthrall as does war.

While, in calling war "the father of all things," Heraclitus exaggerated—though not, one suspects, by much if one recalls ancient Greece—he also embellished a truth. War is *productive* destructiveness,

---

* Current constructions of the human body are hypermasculinized. Women are enjoined to become tough, hard, thin; to radiate intimidating strength; to exclude their own softness, derided as "flab." For many contemporary women, body fat has acquired the organizing status of dirt/ impurity. No analogous softening of the male body is culturally sanctioned. Jane Fonda's *Workout Book(s)* thus far fail to explain the meaning of all this.

† Unless they are named Elizabeth I, Margaret Thatcher, and a few others—exceptions who prove the rule.

not only in the sense that it shifts boundaries, defines states, alters balances of power (*that* we understand)—but in a more profound sense. War creates *the* people. War produces power, individual and collective.* War is the cultural property of peoples, a system of signs that we read without much effort because they have become so familiar to us.

Thus we come to a first vexation: How have we read women's violence as a collective activity altogether out of the picture or as an anomalous appearance each time it appears? The violence of female groups is a sign that signifies formlessness, dis-order, breakdown, mis-rule.

Female Group Violence

On 7 July 1677, Robert Roules of Marblehead, mariner, "aged thirty years or thereabouts," offered a deposition under oath concerning a fearful incident to which he had been witness. It was a time of tension and fear in New England. King Philip's War had just come to an end in southern New England, but the coastal settlements of Maine were being ravaged by a struggle between the Indians and the English settlers. English settlements as far east as Casco Bay had been laid waste, and Mugg Heigon, a leader of the Sokokis, proposed to the Canibas tribe a startling plan: to burn Boston. The Indians would seize English fishing and trading vessels, take them over, and sail into Boston Harbor, taking the city by surprise. During the summer of 1677, twenty vessels were seized, most of them manned by Marbleheaders. Each vessel had a five- or six-man crew. These men had either been taken hostage or killed. The families of Marbleheaders had no word on their fate, prayed for their return, but tensely anticipated the worst. "Every person, almost, in the two colonies [Plymouth and Massachusetts], had lost a relation or near friend, and the people in general were exasperated."

On 15 July 1677, a ketch recaptured from the Indians sailed into Marblehead harbor with two Indian prisoners to be bound over for

---

* Though some individual and collective identities are more power-ful than others.

trial in Boston. The Indians were in bonds. It was the Sabbath. The women of Marblehead were in the meetinghouse. As word spread that a ketch, manned by settlers, was in the harbor with Indian prisoners aboard, the women poured out of the meetinghouse, other townspeople gathered, emotions ran high. The Marbleheaders hailed the sailors but demanded to know why the Indians had not been killed. Told that the Indians were to be bound over, first, to the local constable, then to Boston to be "answerable to the court," the good citizens were not placated. Roules's deposition tells what happened next:

> Being on shore, the whole town flocked around them, beginning at first to insult them, and soon after, the women surrounded them, drove us by force from them, (we escaping at no little peril,) and laid violent hands upon the captives, some stoning us in the meantime, because we would protect them, others seizing them by the hair, got full possession of them, nor was there any way left by which we could rescue them. Then with stones, billets of wood, and what else they might, they made an end of these Indians. We were kept at such distance that we could not see them till they were dead, and then we found them with their heads off and gone, and their flesh in a manner pulled from their bones. And such was the tumultation these women made, that for my life I could not tell who these women were, or the names of any of them. They cried out and said, if the Indians had been carried to Boston, that would have been the end of it, and they would have been set at liberty; but said they, if there had been forty of the best Indians in the country here, they would have killed them all, though they should be hanged for it. They suffered neither constable nor mandrake, nor any other person to come near them, until they had finished their bloody purpose.[3]

Not a pretty picture—but a compelling one. First, because it has eruptive power, flying in the face of simplistic divisions between violent men, pacific women. Second, because it describes chillingly the *form-lessness* of collective female violence; it points to the fact that female violence as narrative often appears as an out-of-control mob, a crowd, a food riot, usually of lower-class composition.* But the good women of Marblehead were no Parisian *canaille*; they were solid citizens who

---

* Natalie Davis goes over this ground briefly, highlighting female food rioters, among other examples, and posing the question why the categorical distinction between men and women in regard to violence despite a long list of female infractions.[4]

murdered and ripped apart two helpless men just after Sunday meeting. Poor Roules clearly feared for his life. His astonishment and horror seeps through his deposition.

He, and the other men, were carrying out their coastal war in an *orderly* way: one takes prisoners rather than dispatching them outright. Roules saw himself as a protector of the Indians *against* the female mob. He would take them to court; they might even be strategically advantageous as live hostages. But the women wanted revenge for (imagined) foul deeds. After the fact, presumably, all got off scot-free. And this, too, fits, once one has absorbed the story's initial shock, for female violence slides off the edge of a divide between bounded/unbounded activity, falling into the unstructured, chaotic, marginal as something disintegrative, anomalous, threatening. Roules recognized nobody; he could describe no one. Perhaps he was too busy looking away.

The "female" has occupied a symbolic and social site deemed potentially uncontrollable: "it is felt that women can loose mindless destruction and violence on the world about her."[5] As the "lustful, disordered and unstable sex," women have not been held wholly accountable "for what they did," writes Natalie Davis.[6] Women have often been let "off more lightly" than men for identical conduct, even when that conduct is violent. The female self in the West has been a latecomer to the full force of juridically constituted legal subjects, the world of law-giving and rule-following males. Because such juridically constituted subjects have been what politics in the modern age presupposed, women have been held at a remove from political accountability.

Female violence, it followed, was an aberration, an eruption of not wholly disciplined subjects, partial outlaws. Not being politically constituted, women are not politically accountable.[7] Male violence could be *moralized* as a structured activity—war—and thus be depersonalized and idealized. Female violence, however, brooked no good. It was overpersonalized and vindictive: behold the "vengeful women of Marblehead."

Violent deeds may be done, but they will go unnarrated, not becoming part of a given society's self-definition. The women of Marblehead are *not* exemplary. Their mob action is no tale of edifying defiance but a tale of shared, merciless gore. Contrast the Boston Tea Party, also a mob action, carried out under cover of darkness. This event is given revolutionary significance, exalted into our reigning narrative. Not all male mob action is culturally sanctioned, of course, the lynch mobs of the American West or the Jim Crow South being now universally decried. But collective male action can be moralized, can take place *within* the boundaries of the culturally sanctioned.

Outside a horizon fused with the story of war/politics, female violence is what happens when politics breaks down into riots, revolutions, or anarchy: when things are out of control. When women transgress, derealizing themselves as pacific and pacified beings, the options are limited. They can either leap back across the divide as quickly as possible, even if it means shaming and punishment, or hover in the marginal interstices of cultural life. Men who cross boundaries in the matter of violence have historically had wider options. There is the possibility of political sanction, as in the Boston Tea Party. But another transfiguration appears on the horizon of the imaginable: the male warrior who operates within a code, and "on the margin of the code, or even beyond it," is granted or "appropriates the right to pardon, to break through the mechanisms of hard justice, in short, the right to introduce some flexibility into the strictly determined course of human relations: to pave the way for humanity."[8] He can break the sword and rescue *in extremis*. He can restore order, including the order he had violated. He can forgive. We have deeded him that prerogative.

Marguerite Duras, in her memoir of the Second World War as a member of the French Resistance, tells a tale without mercy.* At a center in Paris where prisoners freed from Nazi camps got processed, or their deaths confirmed,

> [a] priest brought a German orphan back to the center. He held him by the hand, was proud of him, showed him off, explained how he'd found him and that it wasn't the poor child's fault. The women [Frenchwomen anxious for their dislocated loved ones] looked askance at him. He was arrogating to himself the right to forgive, to absolve, already. He wasn't returning from any suffering, any waiting. He was taking the liberty of exercising the right to forgive and absolve there and then, right away, without any knowledge of the hatred that filled everyone, a hatred terrible yet pleasant, consoling, like a belief in God. So what was he talking about?

* Though she wrote her memoir in 1944, Duras waited over forty years to publish it.

Never has a priest seemed so incongruous. The women looked away, they spat upon the beaming smile of mercy and light. They ignored the child. A total split, with on the one side, the solid, uncompromising front of women, and on the other just the one man, who was right but in a language the women didn't understand.[9]*

We gaze, as from a vast distance, at the densely entwined filiations of a great mystery.

War lies outside the structure of experience for much of the educated, middle-class audience I imagine for this book. This holds for men and women who are my age or somewhat younger, who entered their twenties in the 1960s and for everyone younger than we are. Our parents remember the Second World War, but we do not. Korea is a blur. And fighting in *the* war (Vietnam) is what others did—men from the inner cities and the country, the ghettos and small towns; not college kids, not many anyway.†

If this is where we are *at*, where have we *been*? I have stated that war's destruction brings into being a gallery of particular male and female identities that we tend to compact into two—soldiers on the battle front, women on the home front—but this reduction is a rhetorical amputation that excises many alternatives, male and female. In the discussion to follow, I will enframe the Ferocious Few, the Noncombatant Many, and the Aggressive Mother as female identities. Their male counterparts (though not precise analogues)—the Militant Many, the Pacific Few, and the Compassionate Warrior—make their appearance in chapter 6. (Some of these titles are ironic; some not.)

The Ferocious Few

I have but the body of a weak and feeble woman; but I have the heart of a king, and of a king of England, too; and think foul scorn that Parma or Spain, or any prince of Europe, should dare to invade the borders of

---

* The priest, a man who has forsworn use of his sexual instrument and is forbidden to shed blood, a pacified man, has not *earned* the right to forgive: the women are merciless in endorsing this unstated cultural rule.

† Some figures from Reston's *Sherman's March and Vietnam*: "Of the twenty-six million draft-age men from 1964 to 1973, sixteen million never served. . . . Only 8,800 men went to jail in the Vietnam era for conscientious resistance."[10]

my realms: to which, rather than any dishonor should grow by me, I myself will take up arms; I myself will be your general, judge, and rewarder of everyone one of your virtues in the field. . . . By your obedience to my general, by your concord in the camp, and by your valor in the field, we shall shortly have a famous victory over the enemies of my God, of my kingdom, and of my people.

—QUEEN ELIZABETH I, on the eve of the defeat
of the Spanish Armada (1588)

Ancient goddesses of war hold little allure for us moderns. They represent women in ways to which we are unaccustomed. When we think of a war deity it is likely to be Mars, god of battle, rather than Minerva, goddess of war *and* wisdom. She represents strategic skill and calm victory. But he is a war lover who takes pleasure in conquest and crows in victory. The Mars/Minerva divide on war and war fighting puts pressure on my suggestion that female violence is represented as disorderly, male as rule-bound. The opposite seems to be true in this case, with Mars more likely to run amuck; Minerva, to counsel caution.* But they precede that hardening of the categories I have already traced.

So, of course, does Plutarch. When he penned the third volume of *Moralia,* he devoted a large section to "Bravery of Women" as part of an inquiry into "similarity and . . . difference between the virtues of men and of women." The "worthy deeds" he recounted narrow the gap, showing women "possessed of fierce and savage spirit," mounting walls, bringing up stones and missiles, "exhorting and importuning the fighting men until, finally, by their vigorous defence . . . they repulsed Philip [of Macedon]."[11] This is a story of the women of Chios. Plutarch went on to describe the women of Argos taking up arms and manning the battlements "so that the enemy were amazed." Those women who fell in battle were "buried close by the Argive Road, and to the survivors they granted the privilege of erecting a statue of Ares as a memorial of their surpassing valour." "Even to this day"—Plutarch's time—this event was commemorated with the "Festival of Impudence" during which the women dressed "in men's shirts and cloaks,

---

* The gods and goddesses of ancient times and places are not my central concern, though everything I have read suggests that all the mythologies include deities that preside over war in its various aspects. In Scandinavian mythology, for example, the Valkyrie are any of the twelve war *maidens* who hover over the battlefields and conduct the fallen warriors of *their* choice to Valhalla. (*Val* denotes the slain in battle; *kyra,* the chooser.)

and the men in women's robes and veils." That a ceremony of reversal serves as the cultural remembrance of the women's valor indicates that, although their battle heroics were a departure from standard expectations, such departures *were* possible.

Reversals and ceremonies of reversal are counterpoint to the classical norm, rites of passage that admitted boys to warrioring, girls to preparation for marriage. These two male/female identities were honored by the Spartans, according to Plutarch, for the only dead the Spartans "named"—their names were inscribed on tombstones—were men who had died in war and women who had succumbed in childbirth. In Athens, too, death was anonymous on the funerary *relievos* with the exception of the soldier and the childbearing woman. In death, a kind of symmetry of honor was summoned for those who have not crossed but instead fulfilled their culture's honored callings.[12]

We know women can be brave but doubt they can be ruthless. We know those made of sterner stuff will defend themselves and their children in the final redoubt, the home/land itself, but doubt women will march out to a nation's defense. We know women have been in uniform, but think of auxiliary services, support, noncombat duties. We can accept female spies, for that is a sexualized and manipulative activity given our Mata Hari–dominated image of it. We think rarely of women who have *actually* fought, who have signed up by disguising themselves as men or volunteering their services to resistance and guerrilla movements; and these, too, get slotted as exceptions that prove *the* rule.

The woman fighter is, for us, an identity *in extremis*, not an expectation. Joan of Arc proves this truism through her challenge to it, as her uniqueness as myth and legend in Western history shows. Joan gripped my childhood imagination, but as much for her brave martyrdom as for her warrioring. I suspect it was and has been so for many who have cherished her story in the nearly six centuries since her death. She didn't enter the sainthood until this century. Her martyrdom figured centrally. So did her virginity. She may have donned male garb, but she was a *pure* woman whose violence, or leadership of violence, was sanctified officially once others granted her voices the epistemological privilege she gave them.

Cross-dressing is a common characteristic of the Ferocious Few.

Joan underwent a complete reversal. Elizabeth I, as war queen, decked herself out in some of the regalia and emblems of war. Calamity Jane defied sensibilities on the American frontier by refusing to ride side-saddle "like a lady" and by dressing like a man (see photo section, #3). Less well known women outlaws opted for cross-dressing when they were plying their trade as bank and train robbers and cattle rustlers or biding their time as cowpunchers.* Women cross-dressers are not identical to women in uniform. For a woman to don the uniform of an organized military machine doesn't titillate us with possibilities of a *reversal*, a dramatic rupture. Instead, she becomes *uni-form*. The reversals of Joan of Arc or Calamity Jane are semiotic surprises—unexpected, doing violence to normal anticipations, inviting angry or awed reactions.

A WOMAN FIGHTER IN THE AMERICAN REVOLUTION

Female fighters have surfaced throughout our history as stories of private trangression—an *honored* boundary crossing rather than a repudiated misstep. This is a Revolutionary War story of such a woman fighter.

Though Deborah Samson is "not comparable, certainly, to the prophetess in whom France triumphed" (Joan is drawn upon for mythic, structural legitimation by the teller of Deborah's tale), it "cannot be denied that this romantic girl exhibited something of the same spirit with the lowly herdsmaid. ... There is something moving and interesting in the aspect of enthusiasm fostered in her secret soul." Deborah, who hailed from a poor Plymouth, Massachusetts, family saw the war as an occasion to quit her home for the battlefield. Her motivations were of the noblest, according to her hagiographer, there being "no reason to believe that any consideration foreign to the purest patriotism, impelled her to the resolution of assuming male attire, and enlisting in the army."[14]

Deborah named herself Robert Shirtliffe when she presented herself, in October 1778, "as a young man anxious to join ... efforts ... to oppose the common enemy." Her conduct was exemplary. She vol-

* Jack Weston argues that women cowhands, a minority among the female population but an important group, and black and Hispanic cowboys have disappeared from our standard narrative, the classic western movie which has shaped the popular culture of America—and shaped the world's views of the nation as well.[13]

unteered for hazardous duty. Her first injury, a superficial sword cut, was followed by one far more serious when she was shot through the shoulder. Realizing the wound was a bad one, her "first emotion" was a "sickening terror at the probability that her sex would be discovered. She felt that death on the battlefield were preferable to the shame that would overwhelm her." Deborah eventually married and lived to a ripe old age. Although she had been unmasked (re-sexed as a female), Washington himself discharged her honorably and mentioned not a word concerning her anomalous presence. His silence seems to have spared her humiliation beyond measure as, by her account later, "one word from him would have crushed me to the earth."[15]*

HELL HATH NO FURY: A STORY OF FEMALE REVENGE

> For a while I could bear a grudge against them, it was quite plain and clear, I wanted to massacre all of them, the whole population of Germany, wipe them off the face of the earth, make it impossible for it to happen again.
>
> —MARGUERITE DURAS, *War*

Private revenge is valorized in the first permanent memorial honoring a woman in America, a statue to one Hannah Duston, a thirty-nine-year-old mother of twelve who, aided by her nurse and a teenaged boy, massacred ten sleeping Indians as the three white hostages were being taken to slavery in Canada (see photo section, #7).† Hannah had seen her newborn baby bashed to death against a tree by her captors, and assumed that her husband and eleven remaining children had perished horribly. The victims of her revenge—described as "fem-

---

* Rita Mae Brown recently, in a Civil War novel, revived the story of the female warrior. Geneva Chatfield, a skilled horsewoman, loses her husband to the war the day after their wedding. Not content to usher him off to war, she becomes "Jimmy Chatfield" and enlists in the Confederate cavalry to be near her husband. He turns out not to be much of a man after all: too squeamish for war. She, however, knows how to be a man and acts accordingly. The book's last chapter is a variation on a classical theme, as grandmother Geneva tells granddaughter Laura what she did in the war: war was "exciting," but it was also horrible. "And there were more like her, hundreds. More women fought on our side than theirs." This is clearly a point of pride for Brown's character and for Brown.[16]

† James Axtell tells the story of how many hostages "went Indian," becoming White Indians and preferring, when they were "rescued," to remain with their adopted Indian kin than to return to "civilization."[17]

inine valor" in a story of her exploits—included two Indian braves, their wives, and their six children.

After bashing the sleeping Indians to death, Hannah and her companions took the scalps of "the ten wretches." She subsequently petitioned the legislature—women had no official legal standing—and was granted a Massachusetts government bounty of fifty pounds per Indian scalp, although the bounty law had been repealed the year before her derring-do. The statue, erected in 1874, portrays Hannah Duston "with a hatchet in her right hand, and a bundle of scalps gripped in her left fist," her flowing garment sliding perilously off one shoulder, nearly exposing a breast. The sculptor must have had Lady Liberty in mind.[18]*

FEMALE FIGHTERS IN GROUPS: RESISTANCE, REGULARS, TERRORISTS

> At that time it was clear that each Nazi I killed, each bomb I helped to explode, shortened the length of the war and saved the lives of all women and children.... I never asked myself if the soldier or SS man I killed had a wife or children. I never thought about it.
> —MARISA MASU, Italian Resistance fighter
> Second World War

Sometimes the few become many: in resistance groups, in the armed forces, even as terrorists. Nobody knows how many women participated actively in resistance movements in Nazi-occupied Europe. One writer claims "tens of thousands" in the French Resistance alone, operating as "couriers, spies, nurses, saboteurs and armed fighters."[19] And no more than men were women exempt from the orgies of revenge that followed on the heels of liberation, especially in France. Marguerite Duras describes the brutal interrogation of an informer by herself— cast in the account as a woman named Thérèse—and several of her Resistance comrades. The man they had at their mercy was taunted and then tormented: "Extract the truth this swine has in his gullet. Truth, justice. For what? To kill him? What's the use? But it's not just him. It doesn't concern him. It's just so as to find out. Beat him till

---

* The role of Indian women in the torture and mutilation of white captives or dead soldiers horrified and titillated the imaginations of whites, inspiring, sometimes going beyond, mimesis.

he ejaculates his truth." "Thérèse is me," Duras wrote forty years later, "The person who tortures the informer is me." And Thérèse is justice: the swift justice of retribution, torture and "liquidation."[20]*

Should the few become not merely many but more—and yet more? Join the ranks of male combatants? In France in the Second World War, "*in the tradition of Joan of Arc*, women led partisan units into battle. . . . During the liberation of Paris women fought in the streets with men" (italics added).[21] Some "regulars," members of the armed services, recalling events forty years later, remain vexed by restrictions on what they would and could not do—for example, women pilots for Britain's Air Transport Auxiliary—as others detail wartime camaraderie and equality with men. "I was considered a comrade, just the same as them," reports one French Resistance fighter. But she adds that this was true only in the Resistance: the professional army never accepted a woman as equal.

Women have described their wartime activities as personally *liberating* despite pervasive fears and almost paralyzing anxieties. None regrets her choice to fight or to be in the thick of the fighting. They would, to the woman, do it all over again. But they hope no one else will have to in the future. All report gaining respect, in the words of a Soviet woman fighter, for "men—soldiers—born not out of idolatry from afar, but out of sharing this with them, exposed to their weaknesses, seeing how they coped, and showed more human sides. . . . They cried, they were frightened, they were upset about killing." So were the women. But they did what they had made their duty.[22]†

Today the United States Marines are training women for combat despite the no-combat rule because women "can be assigned to support units that might unexpectedly come under attack and since there is always a danger of terrorist activities." Because women Marines are not assigned to units that are likely to be direct combat units, they are

---

* Duras's account is bitter reading, a reminder of the way one comes to resemble constructions of "the enemy" if the struggle is fierce and brutal and seemingly without end.

† Marshal Tito had high praise for women in the ranks of his Yugoslav Partisans. Their presence "confirmed the quest of women for emancipation" and "inspired the men to heroism." Often "the women were braver than the men, perhaps because the very fact of joining the army and the revolution constituted a greater turning point for them." Tito's rhetoric was a bit high-pitched, clearly constructed for reaffirming the glories of his revolution as much as honoring women Partisans.[23]

not instructed in "bayonets, offensive combat formations, offensive techniques of fire," and a few other skills. Women comprise about 5 percent of the active-duty strength of the Marine Corps.

More interesting than almost-Marines or future conflict is the little-known or remembered story of Soviet women in combat in the Second World War. Soviet women formed the only regular female combat forces during the war, serving as snipers, machinegunners, artillery women, and tank women. Their peak strength "was reached at the end of 1943, at which time it was estimated at 800,000 to 1,000,000 or 8% of the total number of military personnel."[24]*

Soviet women formed three women's air regiments and participated in minesweeping actions. Nadya Popova, a Soviet bomber pilot, has recounted her wartime experience in the language of pure war, the classic language of force Clausewitz would recognize and endorse: "They were destroying us and we were destroying them. . . . That is the logic of war. . . . I killed many men, but I stayed alive. I was bombing the enemy. . . . War requires the ability to kill, among other skills. But I don't think you should equate killing with cruelty. I think the risks we took and the sacrifices we made for each other made us kinder rather than cruel."† Despite this unusual—given the numbers of women involved and the tasks to which they were assigned—experience with women as wartime combatants in regular forces, the Soviets have returned to the standard model, with women designated as noncombatants and vastly outnumbered by the men, fewer than 10,000 by estimate. Just as the Greek term for *courage* is elided to the word for *man*, in Russian *bravery* is by definition masculine. Pointing this out, Shelley Saywell has noted that all the women fighters she interviewed said that women "do not belong in combat," and that they took up arms only because Russia faced certain destruction.[26]

Currently, the most Ferocious Few are women terrorists. From the Baader–Meinhof gang in Germany (Meinhof's first name was Ulrike), to Italy's Red Brigades, to America's Weatherpersons (groups most active in the 1970s), a handful of women have found a home inside the world of terror.‡ Certainly when most of us think "terrorist" we

---

* Far more Soviet women served as doctors, medical orderlies, typists, cooks, and so on, than in combat.[25]

† Ludmilla Pavlichenko, a sniper, was credited with killing 309 Germans.

‡ It seems worth mentioning again that our "terrorism" is often somebody else's "freedom

do not *see* "woman." The perpetrators, or the alleged perpetrators, who have flashed across our television screens in recent years are pretty much a youthful male lot. But women are there, on the front lines and in the support base. Still, when a woman gets accused of an unusually dirty deed, we are shocked—as was I on 3 April 1986 when, sitting in an airport lounge, I saw an "ABC News Brief" report the explosion in a TWA 727, bound for Athens, that sent four human beings plummeting to their deaths: a middle-aged man, a young woman, her baby, and the baby's grandmother. Anchorman Peter Jennings was, as I noted in my journal, unusually somber as he announced that the person suspected of putting the plastic explosive in the plane was "a *woman*"; he looked stunned as he stressed the word. The waiting passengers were clearly disturbed, and as I watched them, I thought, "Equal opportunity terrorism."

The startling reversals of the Ferocious Few—even when the few is a respectable many—are not transformative, or seem not to have been in the West, given the force of our cultural interdiction on serious reflection concerning women's various relationships to life giving and life taking. Women warriors, like their male counterparts, see their violent actions as a form of defense, preservation, life saving. Remember the words of Marisa Masu in the epigraph to this section: "Each Nazi I killed . . . shortened the length of the war and saved the lives of all women and children."[27] Gary Cooper's Sergeant York, in the 1941 movie, says the same thing when he is asked about his record-breaking German sniping and prisoner taking. He did it to try to hasten the end of the thing and to save lives. The sooner it stops, the sooner the killing stops. It remains easier for us to sanction the self-definition of one who kills to bring an end to the killing, if the first-person narrator is a man and not a woman. Many, of course, will see in York's simplicity and Masu's pithy formula only base rationalizations for their own violence. This attitude I believe to be wrong, a way of dismissing Masu and Sergeant York and all the many thousands who have acted on similar principles, and of failing, thereby, to do justice to the problem

---

fighting." Be that as it may, persons who plant bombs in airplanes or railway stations, or mow down passengers in an airport who are preparing to fly off for a Christmas holiday, are properly terrorists by any moral reckoning that makes sense to me.

of "dirty hands," to the necessary moral ambiguity of *any* action in and on the world.

Functioning as compensatory fantasy or unattainable ideal, tales of women warriors and fighters are easily buried by standard repetitions. Framed by the dominant narrative of bellicose men/pacific women, our reflections often lack sufficient force to break out, remaining at the level of fragile intimations. As representations, the Ferocious Few are routinely eclipsed by the enormous shadow cast as the Noncombatant Many step into the light.

The Noncombatant Many

While my conscience told me the war was a terrible thing, bloodshed and misery, there was excitement in the air. I had just left college and was working as a substitute teacher. Life was fairly dull. Suddenly, single women were of tremendous importance. It was hammered at us through the newspapers and magazines and on the radio. We were needed at USO, to dance with the soldiers.

—DELLIE HAHNE in Studs Terkel, *The Good War:
An Oral History of World War Two*

There is general agreement that the answer to the great potential gap between the need for and supply of industrial manpower in America is *womanpower*. So it has been abroad in countries long at war. So it is bound to be, if we must send millions of young men to a score of fronts and at the same time serve at home as the arsenal of democracy. . . . Patriotic emotions are no substitute for intelligence but rather a driving force for the sustained exercise of intelligence. The tasks to which women are assigned must be studied in relation to women's aptitudes, attitudes, and capacity just as any job at any time must be analyzed to be adequately manned. . . . The shift of women from home to factory is not merely an economic change in the labor market but a critical change in social arrangements.

—HELEN BAKER, *Women in War Industries*

As it turned out, my stories had nothing to do with my banishment. I was being thrown out on orders of Lieutenant General Walton H. Walker because I was a female and because "there are no facilities for ladies at the front."*

—MARGUERITE HIGGINS, *War in Korea: The Report of a Woman Combat Correspondent*

Women are the designated noncombatants in modern nation states.† Before modernity and the determined honing of normalized distinctions between male and female activities and spheres of social life, there was no call for a division between all males of a certain age as potential combatants and all females of all ages as noncombatants. The very presumption of an absolute schism between war and social life is an essential feature of bourgeois society with no critical purchase in societies organized differently, whether in pre-modern Europe or outside Europe.[28] The medieval warrior class was a small minority who devoted themselves full-time to fighting sanctified, in part, through the feminization of chivalric discourse. Women were witnesses to male bravery and prowess, on the field of battle (the tournament) and the field of love.

Being the purveyors of war trophies and spectators to male bravery is a task that drew women closer to the circle of actual combat in many pre-state societies. In the Germanic tribes described by Tacitus and Plutarch, sometimes women were depicted as charging with weapons. But more often they were the sacred witnesses to male bravery: "the soldier brings his wounds to his mother and wife."[29]

American Indian tribes were similarly patterned. Tribeswomen were sometimes designated torturers of captured hostages, clearly a violent activity. Women sang songs of war to exhort husbands and sons and were the "sacred witness" to the exploits of male warriors and hunters. In *Son of the Morning Star*, Evan Connell describes a young Oglala Sioux boy of thirteen who, having taken his first scalp, not without difficulty, brings it to his mother, "and when his mother saw the trophy she uttered a shrill tremolo in his honor."‡ Indian women customarily

---

* She got the order overturned on personal appeal to General MacArthur.

† I am not including women peace activists in this discussion. Like the Ferocious Few they, too, depart from the norm, going in the other direction by trying to *pacify* all of society—an aim that, if fulfilled, would leave us *defenseless*, many of their fellow citizens, male and female, insist. See chapter 7 for women's peace politics.

‡ Connell's book is extraordinary, a richly detailed account that makes no attempt to mold Custer and his last stand and the Little Bighorn River into a standard narrative. The debacle

prepared raw scalps for exhibit; they stripped, robbed, and mutilated corpses of the "blue-coats," killing those "who still twitched." And they practiced ritual mutilation as a sign of empathic grief after a battle, gashing themselves "with shards of flint and many of them amputated part of a finger."[30]* Indian women were not warriors but neither were they noncombatants. They were molded to a structure of experience of which war was an inescapable part, as natural to the Oglala Sioux, the Cheyennes, the Arapahos, the Kiowas, and all the others as it was to inhabitants of the great fortified cities of the Bronze Age that gave rise to Homer's epics.

The designation of women as collective nonfighters does not congeal as a presumption of social life until diverse peoples become a *population*, a social body that must be disciplined in order that it be more readily mobilized. In my discussion of the violent wrenchings accompanying the implementation of the discourse of armed civic virtue in the French Republic (chapter 2), I noted the *levée en masse* as the first attempt to gear an entire population for war. The modern—that is to say, eighteenth-century—soldier could be mass produced—this is the ideal—unlike the medieval knight who was a custom-built copy with a lifelong apprenticeship, individualized and ritualized, an aesthetic. Writes Michel Foucault:

> Politics, as a technique of internal peace and order, sought to implement the mechanism of the perfect army, of the disciplined mass, of the docile, useful troop. . . . The classical age saw the birth of the great political and military strategy by which nations confronted each other's economic and demographic forces; but it also saw the birth of meticulous military and political tactics by which the control of bodies and individual forces was exercised within states.[31]†

Though not fighters in this scheme of things, women, too, must work for the national army for it embodies the nation-state. The United

---

comes into view slowly, beginning as something of a sideshow whose enormity was not immediately apparent at all. Connell neither deifies nor demonizes Custer, a dozen or more distinct Indian tribes, the U.S. Cavalry, or anybody else; hence, when Connell does interject some of the language of extermination deployed by Custer and others, the effect is doubly powerful.

* The whites were appalled by such practices, the rather fastidious Custer taking aesthetic displeasure in the whole business. Opening up with gatling guns on unarmed Indians was, apparently, more "civilized" by far.

† The dream of peace, of a perfect social *order*, dating from this period is a dream of domestication, of creating docile social bodies.

States entered full-tilt into this world of nationalizing order with the First World War, as I have already observed; but the way had been paved by the progressivist belief in rational science, efficiency, control. Women—and this varies by class, region, education, a cluster of dimensions—with men were subjected to new modes of discipline aimed at turning them out as proper subjects for *this* sort of society. I am overdrawing the extent and depth of social discipline and conscription of the human body as a predictable cog in the machinery of the social apparatus, in order to press the point that a woman's noncombatant status derives from no special virtue located within her; rather, male bodies are more readily militarized. Female bodies have traditionally had purposes incompatible with the imposition of traditional discipline by the army. Women are designated noncombatants because of the part they play in the reproductive process; because women have been linked symbolically to images of succoring nonviolence; because men have had a long history of warrioring and policing. No state leader—or group of administrators who want a smoothly running machine—is going to put his society through the wrenchings that any wholesale attempt to reverse this scheme of things would require.

Shoved further and further into the background in this scenario—if war is a circle surrounded by ever-widening circles—women retreat from the circle immediately surrounding war in the medieval schema to a circle several times removed in modern Europe. Women still encircle: they are needed to keep the whole thing going. Identities carry over as social practices change: armies still have camp followers; a few wives still trek after their married soldier husbands, a practice less and less tolerated. But women are no longer direct spectators and praise bestowers. They no longer tend the wounds of son, husband, father. That is done either by army surgeons—a fate much feared in the eighteenth century—or, after Florence Nightingale's breakthrough during the Crimean nastiness in the 1850s—by *field nurses*. Compatible with noncombatant status, field nursing places women near the arena of danger but not in the thick of things, locating them figuratively in a familiar (family) way in relation to the dead and dying: women succor, soothe, heal, tend, offer solace. They promise a "woman's touch," a remembrance of home, a dream of surcease and comfort. War nurses were "angels of mercy."

The saga of wartime nursing has been told often enough. The salient point, for my purpose, is that nursing has allowed women to get close to that action of which they are not an active part. In the complex logistics of twentieth-century wars, a small number of men are fighters compared with the number in the rear (base camps) needed to sustain them. Nurses are in or near base camps. Under the peculiar circumstances of the Vietnam War, medical evacuation hospitals frequently came under direct fire.

Occasionally a nurse has transcended the quiet heroics of daily, unrelieved tending of the dead and dying, to share a moment of triumph or defeat. Take, for example, the story of Geneviève de Galard, the "Angel of Dien Bien Phu," who, in 1956, elected to stay with the doomed French garrison, besieged and soon to fall to the Viet Minh. Taken captive by the Viet Minh, she concluded eight grueling years in Indochina as a classic pawn of wartime. She did all these things, she says, not so much *for* the men as to be *with* the men. As she has recalled:

> I remember one young soldier landed near me, and rising to his feet he said, "Well, a woman here! I guess this isn't hell after all." It always amazed me to see the spirit of the men, especially the wounded. . . . Something very special happens among men who are thrown together in very dangerous situations. It's a deep kind of friendship. . . . It is remarkable for a woman to experience and share in this kind of thing, and I found it very moving. I was living through such stirring moments, *as a witness to men's courage.*[32] (Italics added.)

The witness of Vietnam nurses, like that of Vietnam vets, is laced with bitterness. "We were just tired and lonely and sick to death of trying to fix the mutilated bodies of young boys," writes Linda van Devanter. Another nurse, in a letter to van Devanter, says: "Vietnam was not a woman's place. My upbringing dictated that war was a man's job. I came from a very sheltered, structured, protected environment." One wonders how, in that case, she stumbled onto Vietnam, but perhaps her images of the Army Nurse were shaped by war movies just as the images of soldiering were shaped by John Wayne.[33] Despite the ongoing trauma of remembrance, the 7,000-plus women who were stationed in Vietnam (4,500 in the Army Medical Corps) recall those days as *exciting, egalitarian, purposeful, hard-working, simpler,* "*real.*"

Women: The Ferocious Few/The Noncombatant Many

War remains a *spectacle*, drawing males and females, into its orbit.

A second form of battle-zone witness compatible with noncombatant designation is the war correspondent. I have mentioned Marguerite Higgins several times. The woman war correspondent came into her own during Vietnam. Though outnumbered, she wasn't a "freak." The ranks of self-defined female witnesses to future wars is assured. Women reporters are placed in "hot spots" all over the world. Such witnesses are increasingly a part of war action as soldiers "play to the camera" as they fight—or, perhaps, as they fight in order to play to the camera.

But the woman we think of when we think of female noncombatants is the wife, sweetheart, mother, separated (if she happens to be an American) by one of two rather substantial bodies of water from the war itself—hence from her loved one. Women of/on the home front are represented, and have represented themselves, in a variety of ways. They have *inhabited* this designated social role exuberantly or reluctantly, excitedly or resentfully. I shall, first, go over a much reworked theme: what happened when men moved by the hundreds of thousands to war fronts, vacating their previous social locations; and women, at least some of the time, moved in on the territory.

Wars are commonly "named" periods of rapid change in social and political conditions. Elizabeth Cady Stanton, Susan B. Anthony, and Matilda Joslyn Gage, in volume 2 of the monumental *History of Woman Suffrage*, describe the civil war as a time when "new challenges of industry were opened to them [women], the value and control of money learned, thought upon political questions compelled, and a desire for their own personal, individual liberty intensified. It created a revolution in woman herself." The authors condemn war, of course, but they praise the wartime sacrifices of men and women; chide men for having brought us to the brink of war and toppled us over time and time again, yet praise the fervent and spontaneous patriotism war engenders; decry war's terrible cost, yet celebrate and legitimate the many "campaigns" fought by women in sanitary commissions, as suppliers, nurses, comforters, and buriers of the dead.[34] This is a tricky position to sustain—lambasting the social forces that have made possible the activities and identities one extols.

Stanton, Anthony, and Gage did not exaggerate the generative force of the Civil War. According to Mary Elizabeth Massey, [it] provided "a springboard from which they [women] leaped beyond the circumscribed women's sphere into arenas heretofore reserved to men."[35] The First World War brought another great leap forward on the standard accounts. Volume 5 of the *History of Woman Suffrage*, dated 1922, fairly struts over the pages, zealously proclaiming women's loyalty, patriotism, and readiness to serve in a variety of detailed capacities "in event of war," placing the more than two million members of the National American Woman Suffrage Association at the behest of government.[36] The government did enlist some women directly into the service as nurses, technicians, yeomen or yeomenettes. More important by far was the coming to life of the social body as the United States entered the First World War.

Much of this story has already been told, so I will offer only a few highlights as a reminder of women's centrality—her essential contribution to the war effort. Women's war activities gave rise to dozens of women's organizations. Some were anti-war. But peace groups were overshadowed by women's organizations such as the American Defense Society, the National Patriotic Relief Society, the National Security League, and a slew of Woman's Commissions for war relief. Women by the thousands threw themselves into wartime endeavors, often justifying their newfound social identities on the grounds that their nurturing, mothering tradition required it as a moral imperative.[37]*

Mobilized as noncombatants, women went forth—from domestic work to munitions factories, from decorum to daring (smoking began in earnest in the First World War period, women started to drink unashamedly in public, and sexual mores took an alarming plunge—on the view of troubled moralists). Small numbers of women broke into skilled labor, though most did not. The outburst of social-service activities generated by the war incited previously quiescent parts of a woman's self; and startling and irresistible, on women's own account of it, was the discovery that one possessed strength one didn't know one had, that one could enter the world and persevere. To see this starkly as a cut-and-dried case of social coercion and manipulation of women by external forces fails to get at the texture and dimensions of

---

* Other countries were similarly pervaded by women activists. Some experts believe, for example, that women in France realized more progress in three years of war than in fifty years of effort in peacetime. Some, however, contest this view—in the French case and others.

the realizations I have in mind. Women's devotion to, sometimes near obsession with, war work cannot be attributed wholly or even primarily to the xenophobic atmosphere pumped up by vigilante hundred-percenters and the Wilson administration in its least savory aspects. On some level, women had been waiting for this movement to arrive; they were ripe for wider public and social efforts. Unless we understand such latent possibilities, appreciation of the demonstrated enthusiasms of noncombatant women in time of war—or certain wars—will forever elude us.

THE EXEMPLARY TALE OF A NONCOMBATANT:
ELEANOR ROOSEVELT

On 14 May 1915, the young wife of a rising political star wrote a letter to a friend who had lamented the war as "most terrible" before going on to celebrate it in words that spoke of feeling proud and happy, of seeing "all the good qualities, mental and physical strength of our nation." She urged caution, reminding her friend of the many "not beautiful" qualities war also brings out that she herself wished "could be wiped from the face of the earth." The "wonderful, fine qualities" were evoked: she named them "self-sacrifice and unselfishness." But there was also much narrowness and sorrow and bigotry and hatred: "This whole war seems to me too terrible."[38]

But a year after the United States's entry into the war found Eleanor Roosevelt writing her mother-in-law to describe her war work ("knitting at the Navy Department work rooms") as demanding: "It is going to mean part of every day now except Sundays taken up at one place or another but that doesn't seem much to do, considering what the soldiers must do." Her biographer tells us the war "gave her a reason acceptable to her conscience to free herself of the social duties that she hated, to concentrate less on her household, and plunge into work that fitted her aptitudes." Eleanor, a bear for duty, was emboldened by the war to *relocate* the arena over which her duties ranged and to redefine the nature and scope of duty itself.[39]

She ceased being "safe and correct"—her words. A later advocate of full citizen mobilization (men and women) in service to the country, in this earlier war she found a field for action unlike any she had previously known. No one could gainsay the "cause." She learned to do things by herself, developed self-reliance, honed her considerable executive abilities. She yearned to go to England, to be closer to the thick of things; but, contenting herself with flat-out war work in her own country, Eleanor wrote of shifts in her priorities and of gaining assurance in her ability. The war made her a more competent person and, she insisted, a "more tolerant" one. Her imprisonment in Society slowly dissolved. "The war," she wrote, "was my emancipation and education."[40] Glimmerings of a loyalty that is, "above all," loyalty to the truth and to oneself came slowly into focus—as a surprising but welcome spinoff of generous duty to her nation. She ran canteens, made clothing, tended the wounded, set up services for soldiers, an exhaustive list of particular tasks. She emerged from the war years a transformed person. Her own woman: an identity she owed to the war's corrosion of the brittle and imprisoning "social self" she had once been.

Ironies abound. A war had liberated her, by her own reckoning; but she later claimed that "it's up to the women" to make the strongest case for peace. Women must constitute themselves as a democratic "vanguard," realize their "power" and tilt it toward peace. They can begin by educating their children into the adventure and excitement of peace—away from the predominant excitement of war.[41] Her argument trailed off at this point, into a register of the qualms of a liberal-humane conscience. More *to* the point is her *faith* that peace (which is undefined save as war's absence) can function in the profoundly unsettling ways of war. But her *experience* was that war had unsettled her and thousands of other women—and all to the good.

We recognize, in the dimensions of a single life, the conundrum suffragists had earlier produced: we condemn war—but, given war, look at everything we have done!

Think [the prose is unmistakably Cady Stanton's; and the time, the Civil War] of the busy hands from the Atlantic to the Pacific, making garments, canning fruits and vegetables, packing boxes, preparing lint and bandages for soldiers at the front; think of the mothers, wives and daughters on the far-off prairies, gathering in the harvests, that their fathers, husbands, brothers, and sons might fight the battles of freedom.... Think of the multitude of delicate, refined women, unused to care and toil, thrown suddenly on their own resources.

"Think," cried Cady Stanton, and then marvel that we have yet to measure, record, immortalize, monumentalize the memory of women such as these who "passed through the civil war trial of fear as did the men, but their brave deeds go unsung."[42]

The problem that neither Cady Stanton nor Eleanor Roosevelt confronted head on is this: if one actually thinks critically rather than believes by faith, the potential of war as a transformative force looms ever larger. Eleanor Roosevelt was in *peace* before the war broke out. And she was trapped, encased in stultifying, inherited rigidities and social obligations she detested. War was productive of the power of her new-found identity. This is a sobering realization (of which more in my concluding chapter).

COLLECTIVE NONCOMBATANT VIRTUE:
"WHAT DID YOU DO IN THE (GOOD) WAR, MOMMY?"

Women never comprised more than 2 percent of the military in the Second World War—at peak strength, 271,600 women served in some branch of the military—but most women did not sign up. The army was man-made.* A staggering 99 percent of American women led private lives: volunteer activity engaged about one fourth; victory gardens sprouted. Despite the popular appeal of Rosie the Riveter, during the war years nine out of ten mothers with children under six were *not* in the labor force.[43] The only industry women entered to find themselves squeezed out at war's end was heavy industry; in all other forms of manufacturing, their numbers did not decline.

Despite a barrage of government propaganda extolling the femininity of women in overalls who carried wrenches and wore sprightly bandannas round their curls, factory work, according to recent studies, was considered an "ideal work plan" for only one woman in eight in 1943. Mostly women were housewives but, as one scholar puts it, the

---

* Women veterans of the Second World War era are understandably bitter about being treated as second-class soldiers and denied various war benefits. Most, though, look back on the war years as a time they would relive if they could—at least according to the many first-person stories appearing with predictable regularity in the journal *Minerva*.

"entire logic of full wartime mobilization depended heavily on the behavior of housewives," who had to comply with rationing, price and wage controls, frustrating scarcities and add-on work in vegetable growing, finding substitutes for meat, canning, learning new ways to cook.[44] Housewives *complied*. Their civic virtue was secure enough to sustain them through these restrictions and frustrations.

If housewives in sufficient numbers had jettisoned government policies by turning to the black market, fomenting food riots, or encouraging tax evasion, the war effort might have been jeopardized.* "Make do" was the housewife's slogan; plucky coping, her metier. Women took the "home front pledge," in their actions if not with their voices. Historian D'Ann Campbell slips into a near-epiphany when she summarizes the virtues of "Heroines of the Homefront," extolling their "flexibility, creativity, and general competence. . . . They repaired, mended, and conserved. They stood in long lines, lugged home their purchases and children, and devised ways to keep their families clothed, healthy, and well fed. . . . They met adversity head-on." The Second World War's "just a housewife" was, she concludes, quite a gal. Perhaps the most poignant testimony to Second World War change is a single sentence from a black domestic who was sprung from domestic work to a high-paying factory job: "Hitler was the one that got us out of the white folks' kitchen."[45]†

The Second World War is enveloped in a roseate glow. *Life* magazine's commemorative issue on the war in 1985 was emblazoned: OH WHAT A GLORIOUS WAR.[47] One feature is a gallery of "Heroes All," listing, with photos, ten men and seven women—all from the zone of danger. Hot Pilot, Super GI, Pacifist, Top Marshal, Aerial Star, Test Pilot (a woman), Samurai, Martyr, Frogman, RAF Leader, Secret Agent (a woman), Field Nurse (a woman), Flying Ace, and a whole women's page: Spy Chief, Rescuer, Liberator, Saboteur. The Second World War is reinscribed, pumped up yet again as the most glorious

---

* This is Campbell's scenario, and it highlights just how deep was support for the war among America's noncombatants.

† On the whole, the Nazis appear to have done rather less well than the United States in mobilizing women's energies for the war effort: "At the local level, particularly in strongly Catholic areas, ordinary people, especially women, refused to relinquish their habits, let alone their faith, to meet the demands of the new order, and were often either led or backed by a priest or a parson in their resistance."[46]

of our wars. But the 1985 version is just that—a version forty years after the events; hence, the women have come into their own. Women fighters are being *recalled* now; women heroes *remembered*. But Cady Stanton's lament that no songs are sung, no statues built, to the sacrificial heroines of the home front is, apparently, a song without end. *Life* hasn't included as a "Hero All" the "Heroic Housewife." She figures—in stories of the home front replete with recipes for Wartime Cake.

"The Real American Heroine" (not a Hero), *Life* dictates, was Mom. "She said goodbye with a smile on her lips and a prayer in her heart," wrote one grateful citizen in homage to "the real heroine of the war— the American Mom. *Behind every fighting man was a woman.* . . . Blue star mothers had sent sons to the service; gold star mothers had lost them to the enemy." A large photo of a smiling, fur-coated and be-hatted, somewhat plump, pleasant-faced woman in her fifties, reaching down to meet the outstretched hands of workers at a Brooklyn ship-yard—the photo is the sign to which the thin column of commentary is addendum—is the champion gold star mother of the Second World War: "Aletta Sullivan, of Waterloo, Iowa, [became] a five-gold-star mother when her sons were killed on a ship sunk off Guadalcanal"* (see photo section, #13). Mrs. Sullivan became a veritable Everymom: "The boys always wrote at the end of their letters, 'Keep your chin up,'" said Mrs. Sullivan. And now's a good time to do just that."[48] Shades of the Spartan mother!

THE AGGRESSIVE MOTHER: "KILL THEM FOR ME"

> Many a combat soldier in World War II was appalled to receive letters from his girl friend or wife, safe at home, demanding to know how many of the enemy he had personally accounted for and often requesting the death of several more as a personal favor for her!
> —J. GLENN GRAY, *The Warriors*

The aggression of Spartan mothers, American Indian tribeswomen, and American women on both sides of the Civil War has already been

---

* This terrible tragedy (what could the Navy have been thinking to put five brothers from a single family on one ship?) became the subject for one of the greatest cinematic tearjerkers inspired by the Second World War—*The Fighting Sullivans* (1944).

noted, as have the bellicist enthusiasms of women of the First World War era in England and the United States. Some women on the home front went beyond the *sacrificial* stance—a resigned preparedness to "let go"—to a martial enthusiasm to "push out." Notions of male heroism being important to the aggressive mother, she would have them exemplified in her husband, her son. Honor comes to her through her son; less directly, through her husband. Women of Britain Say—"GO!," the famous recruiting poster, pictures a noble mother of classic mein, a daughter in her teens clutching her mother's arms as they enfold one another, a tousle-headed lad around age four or five clinging to his sister's shawl, as all three gaze with determined fondness at the passing backs of a troop of marching soldiers. To be sure, the recruiting poster was put out by the government, not the mothers. But the government clearly knew what message would "sell." The poster crystallized female majority sentiment (GO!) and, together with female recruitment efforts, added up to a potent figuration locating men as heroic warriors who offer up their young lives, if need be, for Mother (Country) and Mum.

"Mother" got drafted into propaganda service over and over, in all warring nations. But the most notorious example comes from Great Britain, remembered now because Robert Graves incorporated the full text without comment in *Good-Bye to All That* (1929):

> A Mother's Answer to 'A Common Soldier.' By A Little Mother. A Message to The Pacifists. A Message to the Bereaved. A Message to the Trenches.

That was how it began, and the letter that followed, having appeared in the London *Morning Post* in 1916,[49] was reprinted, by popular demand, in pamphlet form, selling 75,000 copies in less than a week direct from the publishers. The Queen Mother herself was "deeply touched"—the *Ür-Mutter* authorizing the voice of the anonymous "Little Mother."

The "Little Mother" identifies herself as the mother of an only child:

> [The voice of the] mothers of the British race ... demands to be heard, seeing that we play the most important part in the history of the world, for it is we who "mother the men" who have to uphold the honour and traditions not only of our Empire but of the whole civilized world. ... We

women, who demand to be heard, will tolerate no such cry as Peace! Peace! where there is no peace. . . . There is only one temperature for women of the British race, and that is white heat. With those who disgrace their sacred trust of motherhood we have nothing in common. . . . We women pass on the human ammunition of "only sons" to fill up the gaps. . . . We gentle-nurtured, timid sex did not want the war. . . . But the bugle call came. . . . We've fetched our laddie from school, we've put his cap away. . . . *We* have risen to our responsibility. . . . *Women are created for the purpose of giving life, and men to take it.* . . . We shall not flinch one iota. . . . [Should we be bereft, we shall] emerge stronger women to carry on the glorious work our men's memories have handed down to us for now and all eternity.

Yours, etc., A LITTLE MOTHER[50]* (Italics added.)

Baldly proclaiming a sentiment carved on the mangled bodies of others, possibly her own son, "A Little Mother" expresses blood-curdling patriotism coated in vapid and lifeless pieties. Graves could not believe it; nor can we. But eager thousands did, including many mothers who saw themselves refracted through the "Little Mother's" longing for Spartan motherhood. That mothers have been prepared, over the centuries, to sacrifice their sons on the altar of Mars—either wrenchingly, fearfully, reluctantly, forebodingly, or eagerly, stalwartly, proudly—takes us to the heart of another mystery.

* Graves dryly appends, without comment, both the "Letter from a Little Mother" and "Extracts and Press Criticism," celebrating the Letter as among the "grandest things ever written . . . loftiest ideal . . . most sublime sacrifice . . . exquisite."[51] The effect—lodged, as this extraordinary epistle is, within a small masterwork of ironic discourse—is devastating.

6

Men: The Militant Many/
The Pacific Few

It has always been my ideal in war to eliminate all feelings of hatred and
to treat my enemy as an enemy only in battle and to honour him as a
man according to his courage.   —ERNEST JÜNGER, *The Storm of Steel*

This book is to be neither an accusation nor a confession, and least of all
an adventure, for death is not an adventure to those who stand face to
face with it. It will try simply to tell of a generation of men who, even
though they may have escaped its shells, were destroyed by the war.
                                        —ERICH MARIA REMARQUE
                        Prologue to *All Quiet on the Western Front*

Patriotism. There was no patriotism in the trenches. It was too remote a
sentiment, and rejected as fit only for civilians. A new arrival who talked
patriotism would soon be told to cut it out.
                              —ROBERT GRAVES, *Good-Bye to All That*

T HESE DIVERSE VOICES from the First World War signify
a shift in the burden of exploration as I move from women and war
fighting, or in wartime, to men. War is an experience to which men
are not *exterior* as are women. Men have inhabited the world of war,
called up the "dogs of war," conjured with the cataclysm in a way
women have not. This division has played a constitutive role in the

194

structuring of human societies past and present, and survives despite evidence that suggests that women, too, are capable of violence and of sanctioning the violence of others. Because the assumption is that men are to life taking as women are to life giving, it is that ubiquitous chestnut with which I am here concerned. How have men inhabited the world of war? Are men, in fact, eager life takers, aggressors itching for a fight? Is war, then, a cultural métier that they are uniquely habituated to, indeed structured for? Is it, in fact, the case that war *in essence* is the unleashing of individual aggression on a mass scale?

I shall follow the pattern established in the preceding chapter, interweaving past tales and remembrances, offering narrative space to many voices, bringing forward abstractions and details, commentaries and quotes. The prototypical male characters who will appear are the Militant Many, the Pacific Few, and the Compassionate Warrior. War as a narrated event—one important part of war's reality—will be explored through male and female voices. The war story is a form of discourse that differs from the grand tradition of armed civic virtue in the West, although the two touch upon and help to make one another possible. At times, however, the war story may be told in a voice— ironic, disillusioned, bitter—at odds with the presumptions, or pretensions, of modern-day rationalist realists and celebrants of a grand notion of civic virtue alike. Finally, the structure of experience of the warrior/male and the mother/female will be compared. All has never been quiet on the war-discourse front.

The Militant Many

And yet, in spite of it, there grew a compelling fascination. I do not think I exaggerate: for in that fascination lies War's power. Once you have lain in her arms you can admit no other mistress.... No wine gives fiercer intoxication, no drug more vivid exaltation.... Even those who hate her most are prisoners to her spell. —GUY CHAPMAN
(quoted in William S. McFeely, *Grant: A Biography*)

Many veterans who are honest with themselves will admit, I believe, that

the experience of communal effort in battle, even under altered conditions
of modern war, has been a high point in their lives.
　　　　　　　　　　　　　　　　　　—J. GLENN GRAY, *The Warriors*

Images of marauding males for whom war is an opportunity to kill
with immunity prevail in much thinking about war. While the picture
of would-be murderers champing at the bit has been revived in some
current feminist anti-war rhetoric and analysis, it has an old history.
Searching, one can find evidence to shore up this ugly picture, from
the Homeric epics to running-amuck Rambos.* For example, Tzvetan
Todorov writes of the warrior/woman opposition in Aztec society as
playing a structuring role for "the Aztec social image repertoire as a
whole." The warrior is the "male par excellence, for he can administer
death," and the "worst insult ... that can be addressed to a man is to
treat him as a woman." Women, having "internalized this image ...
contribute to the maintenance of the opposition, attacking young men
who have not yet distinguished themselves on the battlefield." Aztec
women goaded young men into battle, demanding that they prove
themselves.[1]† Even among American Indian tribes in which women
played a powerful and honored role (the Iroquois are a good example),
rhetorically the split between warrior/woman structured male and
female identities. To insult a potential or a defeated enemy, one called
him a "woman," and the warrior loath to fight was "womanized."

Culling sources from classical antiquity to modernity, Mary Beard,
in her *Woman as Force in History* (1948), claims that women in previous
epochs in the West were involved in all aspects of war making, including enciting men "to ferocity at the fighting fronts. They accompanied men on marauding expeditions. They fought in the ranks. ...
There was not a type of war in which women did not participate."
Beard recounts Tacitus's horrified descriptions of first-century A.D.
Germanic tribes whose squadrons of soldiers included whole families
and clans: "Close by them, too, are those dearest to them, so that they
hear the shrieks of women, the cries of infants. The women are to

* Although even in Rambo's case, a note of caution must be lodged: he is on a *rescue* mission;
to save Americans missing in action, he must kill. But one gets the feeling that Rambo doesn't
so much kill to rescue as he rescues to kill.
† Todorov suggests that the binary opposition—warriors/women—structured another—
weapons/words; and that women have, in some sense, won "this war" in European culture since
the Renaissance, which "glorified what we might call the feminine side of culture" by celebrating
words rather than weapons, words that transmit the tradition of war: a war of words.[2]

every man the most sacred witness of his bravery—they are his most generous applauders. The soldier brings his wounds to his mother and wife." Plutarch, in the first century B.C., recalled encounters with barbarous hordes "whose fierce women charged with swords and axes, and fell upon their opponents uttering a hideous outcry. . . . When summoned to surrender, they killed their children, slaughtered one another, or hanged themselves to trees."[3]* In their ferocity, or despite it, the warrior/woman opposition remained intact.

Contemporary American society is bubbling with signs that signify men as a militant, bellicose group, from Rambo-style garb donned by high school students (as a way to promote school spirit) to the Houston nightclub where waitresses wear fatigues and a .50-caliber machinegun is part of the decor. The high school kids who turned themselves out as Rambo made light of it, saying, "It's fun to dress wild without getting in trouble." Local Vietnam vets read this behavior differently, one of them saying: "Many of our brothers went to their graves because they believed wars were fought the way John Wayne portrayed them in his movies."[5]†

Military toys have enjoyed a major resurgence. Sales hit the doldrums during the Vietnam years, but purchase of toy guns and rifles doubled between 1979 and 1982, and G.I. Joe, after a seven-year furlough, got pressed back into active duty by Hasbro Industries in 1982, with sales in 1983 nearing $100 million. A trade magazine for the toy industry crowed: "Think military! . . . Given the new military consciousness, it seems only natural that leading toy companies are putting renewed energy into military-related toy lines."[7] The toys, of course, are primarily destined for male, not female, children whether—one suspects—the child has requested a gun, a G.I. Joe, or a miniature Huey, a helicopter widely used in Vietnam. Films in which the warrior is portrayed favorably continue to pack them in at the box office.‡ As I write, in the summer of 1986, the film *Top Gun* is a top draw. It

---

* Dr. Helen Caldicott's current assurance that women are more pacific because they can give birth seems a less universal imperative than she apparently recognizes. It is unlikely wombs have changed; thus we must search for other explanations for Beautiful Souls and women warriors alike.[4]

† The Rambo incident in question occurred in Greenfield, Massachusetts, in October 1985. About 1,200 people, according to the *Washington Post*, were crowding into the Rambo nightclub every night.[6]

‡ I noted two such films—one geared explicitly toward the woman-warrior identity rather than woman/warrior opposition—in the introduction.

features as heroes navy fighter pilots engaging in high-tech dogfights with Soviet MIGs. Our F-14s in the hands of such crack pilots as "Maverick" get the better of the unnamed enemy as audiences cheer. The movie suggests that war can still be an individual contest of skill between men. A more peculiar sign of the current times is the depiction as "Star Warriors" of men who design the "weapons of the future." Most of the weapons with which they conjure operate, once in motion, independent of human artifice. "Battles" are construed as weapons engaging other weapons in a fratricidal conflict. The image "Star War(rior)" itself, of course, comes from a movie fantasy in a curious mimetic relationship.

More bizarre but suggestive nonetheless is the growth of paramilitary training camps in the United States, really prep schools for mercenaries, in which men (a few intrepid women turn up) take lessons in throat cutting, ear removal, and killing, using "sticks, hands, feet, knives, rope and firearms of all descriptions. Also on the curriculum are ambushing, patrolling, rappelling, camouflage, booby traps and a special pre-dawn seminar in torture."[8] There are a dozen or so such camps in the United States, their students preparing either for a Communist invasion they deem inevitable or for fighting communism in other countries as soldiers of fortune. The magazine *Soldier of Fortune* has grown, in ten years, to a circulation of half a million.* "Damn if I don't love riding my Harley and shooting my gun!" exclaimed one student to a reporter.[9] The training itself is violent, and students are frequently injured. By current estimate, some 500,000 *military-style assault guns* are in private hands in the United States, the vast majority, so far as analysts can tell, in the hands of men, not women. Popular items include the Israeli UZI, the Mac-10, the AR-15. The UZI fires 900 rounds per minute; the AR-15, 650; and the Mac-10, 1100.†

In 1986, Ralston Purina company introduced a "Wholesome Multigrain Cereal With Less Sugar Than Most Sweetened Cereals!," "New G.I. Joe Action Stars." The cereal box pictures a large bowl of cereal in the lower left-hand corner and, dominating the right-hand section,

---

* Competitors with *Soldier of Fortune*, for those whose tastes run to violent imagery, include *Gung Ho* and *New Breed*.

† Although it is illegal to purchase fully automatic weapons, semiautomatics are available and can be converted to full automatics.

a fierce comic-book drawing of a clenched-fist, crew-cutted, hulky blond warrior in tight black pants, combat boots, and an army-type shirt jacket with rolled up sleeves. Downing this wholesome cereal is a smiling young man whose mother, her arms around his shoulders, gazes at him lovingly, a toothsome grin on her face, as he turns himself out as tough and bold through G.I. Joe Action Stars. The cereal must be purchased by mothers for consumption by (male) children. Because companies do extensive market research before they introduce new items, the appearance on supermarket shelves of G.I. Joe Action Stars suggests that women are ready to buy, children to clamor for, a militarized breakfast cereal.

One popular image of what war does or makes possible—the unleashing of aggressive impulses, a *regression* to a less civilized state (construed, always, as animalistic or bestial)—lends credence to the truism that the majority of men are or can be drawn enthusiastically into the enterprise of killing. The militant majority image, as a male identity called forth or made explicit by war, is that of a *killer*, an unchained luster after blood. This image is sustained by public-opinion polls that indicate that more women than men fear war, by 2 to 1; that two thirds of men polled but only one half of women supported the United States invasion of Grenada; that more men than women believe we should be "more forceful" in dealing with the Russians even if it leads to war. One analyst, reflecting on the 1980 presidential election, writes that "the difference in the way men and women cast their ballots could be entirely eliminated by controlling for the willingness of men to be more aggressive in foreign policy, even at the risk of war."[10]*

The "drive-discharge" model underwrites war as a time of unleashing of aggressivity. This thesis derives from psychoanalytically informed theorizing, in its professional and pop versions. Freud is the father of this view—as of so many others that we hold about ourselves—as articulated in two essays: "Thoughts for the Times on War and Death," written in 1916 in the midst of war's carnage; and "Why War?," a 1932

---

* These and similar analyses prompted assurances by women leaders and polemicists—Eleanor Smeal, Bella Abzug, among others—that a "gender gap," plus the presence of a woman on the Democratic ticket, would virtually assure the defeat of Ronald Reagan in 1984. Such, of course, was not to be.

exchange with Einstein on how to build peace. The First World War brought keen disillusionment, Freud opined, because "we" (cultured Europeans) thought "we" had evolved beyond the enormous cruelties war had called forth. Freud's argument, briefly, is that in time of war individuals are permitted to engage in actions they would look at with revulsion in time of peace. This happens—or can happen—because the majority of individuals have not undergone a true inner transformation of aggressive impulses (as had Freud and a minority of pacifists who, on Freud's view, genuinely completed the task of internal renunciation and sublimation of aggressive drives).[11] Regressing to an earlier stage, individuals recapitulate, in a sense, the domination by brute violence with which human history began.

Although Freud and all those who share this presumption offer psychological reversion to a kind of bestiality as *the* explanation for the "killer" identity war generates, others who reject the notion as insufficient see moments of such violence in war fighting. J. Glenn Gray, whose *The Warriors* remains the greatest of all reflective books on men and war, notes one of a cluster of impulses war engenders or helps to *make visible* in a stark way—a *delight in destruction*. This delight is *not* so much that of an egoistic warrior giving himself over to blood lust as a collective recognition of war's power and grandeur. Gray writes of astonishment and wonder at displays of awesome force, noting that war offers these human capacities an exercise field *par excellence*. More sinister by far than intimations of the sublime when one finds oneself surrounded by, or enclosed within, an awesome field of force is the discovery in war, by "thousands of youths who never suspected" it, of the presence in themselves of "the mad excitement of destroying." Somberly, Gray reflects: "Men are in one part of their being in love with death, and periods of war in human society represent the dominance of this impulsion."[12] William Broyles, Jr., reinscribing Gray's analysis to help himself understand his own Vietnam experience, writes: "The love of war stems from the union, deep in the core of our being, between sex and destruction, beauty and horror, love and death. War may be the only way in which most men touch the mythic domains in our soul. It is, for men, at some terrible level the closest thing to what childbirth is for women: the initiation into the power of life and death."[13]

Men: The Militant Many/The Pacific Few

Gray would not exempt women from the generic category *men*. Although women do not do the actual killing, they are part of the structure of oppositions that encourages and requires it. Women also share in and help promote one of several images of "the enemy" that invite the worst of war's excesses. Gray identifies these as the *totalitarian* image which sanctions seeing the enemy as the representative of a principle of evil one must stomp out, and invites a crusading ethos of the sort that characterized much of America's war against the Japanese (and theirs against us) in the Second World War; and the *subhuman* image that sees the enemy as a creature who is not human at all, a noxious, abhorrent "species of animal."[14]

In *Sahara* (1943), one of the best Second World War–vintage films, directed by Zoltan Korda, Humphrey Bogart, a tank commander fighting in North Africa (the time is June 1942), describes a downed German pilot who has been taken prisoner: "He's a Nazi. He's like a mad dog." This dehumanization of the German goes hand in hand with the overcoming of national estrangements, as a polyglot crew, including an Italian prisoner of war, a Frenchman, a British Sudanese, an Australian, a South African white, several Brits, and Americans from Brooklyn and Waco, Texas—and points in between—make common cause: "You sure learn things in the army," says the Texan to the Muslim Sudanese, who replies: "We have a lot to learn from one another." War's irony is here depicted—brutal dehumanization, camaraderie despite difference. Soldiers permeated with either the demonic or the subhuman image of the enemy more easily became brutalized than did soldiers who saw the enemy through the lens of professional soldiering (all military men are comrades in arms) and those who came to view the enemy as a human being, like oneself, who is forced to make war against his better will or desire.

Delight in destruction, made possible by viewing the enemy as either demon or beast, does not mark a transformation from man to beast but signals a way of being human—all too human—given the similarity of creative and destructive urges in most individuals. But even in time of war, only a minority of fighting men are completely and continually seized by a delight in destruction. They do not, for the most part, revel in it, as popular views might suggest and as some radical feminist argumentation insists. With the transition in the West from the "warrior" to the "soldier," a different relationship of the man-in-arms to violence emerged—a point I return to after a brief excursus on the male analogue to the female Ferocious Few.

The Pacific Few

Over hill Over dale
We have left our sawdust trail
(and the caissons go rolling along) Split that oak
Slash that pine
Come on, boys, you're doing fine
(and the caissons go rolling along)
For we're all CO's
In our working clothes;
Chopping will make the muscles strong.
So let's sing and play
All the live-long day
While the caissons go rolling along
(and they'll keep rolling)
While the caissons go rolling along.
  —"Warner Rouser," a tune sung by conscientious objectors
    assigned to a forestry camp located near Warner,
    New Hampshire, beginning October 1942

The song above, for all the plucky funning it exudes, speaks to a curious truth: that the pacifist or CO (conscientious objector) often sees himself in mimetic terms as the *militant* analogue of the violent warrior. A major difference in the statuses of warrior and pacifist is the fact that the pacifist is or has been reviled by and estranged from the "majority," male and female, in wartime. This at least has been our history, with Vietnam a somewhat less clear-cut case. Over a six-year period, about 12,000 men in the United States were classified as Second World War conscientious objectors and assigned to 150 campus and special units of the Civilian Public Service. Memoirists of that experience found it an attempt to punish and suppress dissidents who were overwhelmingly religious in orientation and cited religion as the grounds for their refusal to fight in the war, playing out what one of them called "the ever-present tension between the roles of *believer*, subject to the moral dictates of conscience and the teachings of the Church, and *citizen*, subject to the authority of the state."[15] They were, as they saw it, soldiers of God's peace and truth.

I noted the historical peace churches in chapter 4, indicating that they were and have been unable to stem the rush toward collective violence in American history. Modern total war has not only deepened the conviction of such churches; it has widened the pacifist appeal. "Nuclear pacifism" is now the normative doctrine for the Roman Catholic Church and most mainstream Protestant denominations. It remains to be seen whether this stance will have the effect of blocking the construction of the male self as an embodied militant, a soldier, geared toward war fighting; or whether it will enable more men to turn themselves out as militant pacificists, as nonfighters in a "holy war" against violence—the more difficult alternative given the weight of historic example pushing in the other direction.

The pacifist stance, in the modern West, has not been an easy one for men to attain and to sustain. The 4-E CO classification set men up for public abuse. One Second World War CO describes an incident that occurred when his work camp was being transferred from east to west, requiring a train ride across the country:

> The good ladies were out at the railroad stations with candy and food and magazines.... They would offer gifts to the men that were going off to serve their country.... We'd spill out of the cars onto the railroad tracks, the ladies didn't know who the hell was which. [Whether marine recruit or conscientious objectors, both were being transported on the same train.] So we ended up with a lot of these goodies. When word got around that there were some yellowbellies on this train, the ladies would actually go around and yank us by the arm and say, "Are you one of those damn yellowbellies. I want my cookies back."[16]

Despite such abuse, many wartime COs, in common with wartime fighters, have fond memories of their experiences, recalling solidarity in the camps, the forging of tight friendships under the exigencies of the military regime CO camp members were compelled to follow. Historic accounts of pacifism in wartime tend to ignore this irony— and that of pacifist men being decorated for wartime valor. Among the ranks of pacifists are some who refuse to bear arms but will go to war in some other capacity, most often as stretcher bearers and medics. The most decorated British soldier of the First World War was a stretcher bearer, Private W. H. Coltman; and a Second World War CO, a Seventh Day Adventist, won the Congressional Medal of Honor for saving seventy-five wounded men in one battle despite his own grievous injuries.

For the small minority who are absolutely pacifists (Quakers, for example), together with those who are conditional pacifists (refusing to fight *this* war because it fails to meet the criteria for a just war, as any total war must), many more men find their pre-war pacifistic or anti-militarist sentiments and convictions dissolving under the pressure of wartime exigency. Freeman Dyson writes movingly of this transition (as do some of Studs Terkel's respondents in *The Good War*) as he traces his own retreat, during the Second World War, from one moral position to another:

> At the beginning of the war I believed fiercely in the brotherhood of man, called myself a follower of Gandhi, and was morally opposed to all violence. After a year of war I retreated and said, Unfortunately nonviolent resistance against Hitler is impracticable, but I am still morally opposed to bombing. A few years later I said, Unfortunately it seems that bombing is necessary in order to win the war, and so I am willing to go to work for Bomber Command, but I am still morally opposed to bombing cities indiscriminately. After I arrived at Bomber Command I said, Unfortunately it turns out that we are after all bombing cities indiscriminately, but this is morally justified as it is helping to win the war. . . . I had surrendered one moral principle after another.[17]*

Dyson had, in his own view, surrendered to the wartime logic of force.

Just as our cultural proscription on serious and nuanced reflection on women and violence limits our capacity to understand what *is* or *has been* and strangles our ability to play with alternatives, so the presumption that men = violence and aggression constrains the emergence of other options. These options, while not so severely constrained as those hemming in talk of women and violence, are nonetheless pow-

---

* Max Hastings tells the story of a pilot who, thirty-five years afterward, still had nightmares about the destruction he had unleashed over Germany. His brooding led him to change jobs and to teach mentally handicapped children "as a kind of restitution," but he remained "haunted by the war."[18]

erful. Ironically, men who have "been through it" have much greater cultural sanction to talk about it and suggest "never again," than do men who have never faced the dilemmas of wartime fighting or been through the liminal experience of a war front.*

It was not always thus. "Once upon a time"—there were many more culturally endorsed ways to be a man, to inhabit a male body and to use that body to purposes other than, or at odds with, sanctioned violence. "Thus, religion offered models for masculine behavior different from that of the warrior, however just and brave," writes Natalie Davis. Class differences also played a role, with the aristocrat being one imbued with strong notions of honor that might require fighting, and converts to the "holy wars" I described in chapter 4 eager to slay the demonic foe.

But many men dragooned into the army deserted as soon as they could, or avoided actual combat as long as they could. This is, as Davis points out, the male as "Trickster . . . who uses his wits to survive"; and he is limned as the prototypical picaresque hero, Simplicissimus, "who invented a listening device so powerful that he could hear the enemy approaching three hours in advance and avoid encounter betimes, and who devised shoes that could be worn backward so that the enemy would not know in which direction he had escaped."[19] The priest is disarmed, as are women; the Trickster is crafty, as was Eve— challenging later convictions that construe the relation of men and women to violence in simplistic ways, making of the Ferocious and the Pacific Few cultural strangers, oddballs in our midst.

Neither the Militant Many nor the Pacific Few capture the exemplary male identity called forth by war, however; and it is to that I next turn.

THE COMPASSIONATE WARRIOR: WARTIME SACRIFICE

> Our function was not to kill Germans, though that might happen, but to make things easier for men under our command.
> —ROBERT GRAVES, *Good-Bye to All That*

> Comrade, I did not want to kill you. . . . You were only an idea to me

---

* A liminal experience involves boundary crossing, marginality, the uncanny. War fighting is such an experience because men are compelled to forsake the framework of normal life and, in taking up arms against others, to alter drastically their personal identities and social circumstances.

205

before, an abstraction that lived in my mind and called forth its appropriate response. . . . Forgive me, comrade.

—ERICH MARIA REMARQUE
All Quiet on the Western Front

I understand . . . why I jumped hospital that Sunday thirty-five years ago and, in violation of orders, returned to the front and almost certain death. It was an act of love. Those men on the line were my family, my home. They were closer to me than I can say, closer than any friends had been or ever would be. They had never let me down, and I couldn't do it to them.

—WILLIAM MANCHESTER, *Goodbye, Darkness:*
A Memoir of the Pacific War

Man, as designated combatant, presents himself in his most prototypical guise in the West as neither a wholly reluctant warrior—though there is much of that—nor as a bloodthirsty militant—though there is some of that. Rather, by his own account of wartime experience, he constructs himself as one who places highest value not on *killing* but on *dying*—dying for others, to protect them, sacrificing himself so that others might live. This theme, resonating consistently and powerfully, cannot be ignored.

Although his sacrifice is seen by noncombatants, including women, as a sacrifice *for* the nation, for the collective, the soldier himself is far more likely to think and act in terms of his immediate cohort, that small number of men he is actually fighting, and possibly dying, with. Loyalties of soldiers in wartime are to comrades and buddies, not to states and ideologies, or these are the exceptions. Ernst Jünger, one of the least sentimental observers of the First World War, valorizes the distillation of the Fatherland that emerged from the war as a "clearer and brighter essence," as men learned to "stand for a cause and if necessary to fall as befitted men."[20] But Jünger's voice seems rarer than the stark despair of Remarque's protagonist Paul Baumer, for example. Japanese soldiers in the Second World War played out the theme with their own variation—an oath of personal loyalty to the emperor. For those who volunteered to die as kamikaze, this oath was a matter of honor unto death, a sacrificial seal.

Richard Holmes apprises us that of eight medals won by Marines on Peleliu in 1944, "six were awarded to men who covered grenades with their bodies to save their comrades." Five black Marines earned the Medal of Honor in Vietnam: *All five* were killed shielding fellow Marines from exploding enemy grenades.[21] One can put things even more strongly. S. L. A. Marshall, the great American military historian, concluded, as a result of his extensive study of American rifle companies in action in the Pacific and European theaters of the Second World War, that *"fear of killing, rather than fear of being killed, was the most common cause of battle failure."* Moreover, on the average, only 15 percent of these men actually fired their weapons in battle, even when they were very hard pressed by enemy soldiers. Writes Marshall: "Seventy-five percent will not fire or will not persist in firing."[22] Men behave defensively instead, trying to spare themselves and, more importantly, to protect their comrades. Inhabiting what John Keegan calls a wildly unstable environment, soldiers are bound up with their most immediate comrades and think of themselves as "equals within a very tiny group" of some six or seven men.[23]*

J. Glenn Gray, from the soldier's-eye point of view, examines the impulse to self-sacrifice characteristic of warriors who, from compassion, would rather die than kill. He calls the freedom of wartime a *communal freedom* as the "I" passes into a "we," and human longings for community with others find a field for realization. Communal ecstasy explains a willingness to sacrifice and gives dying for others a mystical quality not unlike the one abstractly depicted by Vera Brittain (see page 211) at her safe remove from trench warfare. Because this fits no rationalist image of what human beings are all about and what makes them tick—sacrifice being implacably at odds with self-interest—rationalists must disdain the supramoral act of sacrifice or gloss it over: "Such sacrifice seems hard and heroic to those who have never felt communal ecstasy. In fact, it is not nearly so difficult as many less absolute acts in peacetime and civilian life. . . . It is hardly surprising that few men are capable of dying joyfully as martyrs whereas thousands are capable of self-sacrifice in wartime."[24]†

---

* Because this bonding of the small group has been essential to war fighting, Keegan concludes that, under modern conditions of techno-war, it will be less and less possible to develop the relations and identities necessary to carry out sustained war fighting between groups of men.

† If Gray is right, the impulse to sacrifice I noted in my discussion of the early Christian martyrs may not capture people's energies in the same way any longer because modern states separate dissenters and make them die *alone*. If they could die together—as one part of a larger

LIFE GIVERS/LIFE TAKERS

Remarque's doomed protagonist reflects: "We are soldiers. It is a great brotherhood, which adds something of the good-fellowship of the folk-song, of the feeling of solidarity of convicts, and of the desperate loyalty to one another of men condemned to death, to a condition of life arising out of the midst of danger, out of the tension and forlornness of death."[25] Soldiers in wartime frequently make provisional peace, declaring their own truces on the front, showing one another pity and compassion. Siegfried Sassoon noted a time when, on the Western front, the Germans could undoubtedly see the arduous effort their enemies were engaged in as they worked feverishly to repair the human and tactical damage done by a bomb explosion but "stopped firing: perhaps they felt sorry for us."[26] For each instance of brutality, a deed of sacrificial courage or benign restraint can be proliferated—if one is telling the story *from the ground up*, as a narrative of men's experience rather than as an account of strategic doctrines or grand movements of armies and men seen from a bird's-eye point of view.

The lessons learned by Yoshida Mitsuru, a law student at Tokyo Imperial University in 1943 when he was called up into the imperial Japanese navy, and one of the few survivors of the sinking of the battleship *Yamato*, tells those who have not died in war (as he assumed he would): "Never underestimate how precious life is."[27]* Families of American marines killed in the barracks explosion in Lebanon, on 29 August 1983, reading from letters they had sent home, reported hopes for peace. "He wanted peace," said Carol Losey, mother of twenty-eight-year-old Second Lieutenant Donald Losey, "he [wanted to] help." Others: "I feel proud to be part of a peacekeeping force." "I really want to go home. But deep down in my heart I want to be part of something that turned out good." "I have a job to do. Protect the people."[29] What is remarkable about these expressions is their very unremarkableness. The soldier is one who serves and, if need be, sacrifices for others.

Together with the powerful sacrificial possibilities and idealizations

---

group, in the assurance that group life will continue—then the analogue to sacrificial warrioring would work better.

* The classical samurai warrior code, a strong ethic of sacrifice being its leitmotif, is captured in Yukio Mishima's chilling tale "Patriotism"—a story that presaged Mishima's death, in 1970, by his own hand in an act of ritual *seppuku* (disembowelment).[28]

208

war fighting makes possible and crytallizes as a form of male identity, visible on all battlefields is a category of concerned attachment that Glenn Gray calls "preservative love," an impersonal passion to preserve and to succor, to hold back the annihilation. This concern seems to Gray "maternal" in nature, protective in its unfolding. One finds side by side the paradox of organized mass killing and caring for small, vulnerable, particular things. Men grew flowers and vegetables outside their trenches on the Western front. Adopting animals—stray puppies, even small mice—as pets is a common practice, noted by war correspondents like Ernie Pyle, described by men who have been in war like Gray, Robert Graves, Hemingway. "Waifs and orphans and lost pets have a peculiar claim on the affections of combat soldiers, who lavish upon them unusual care and tenderness," writes Gray.[30] The spectacle of wounded or killed animals, especially horses dragged into the fighting, is scored as particularly abhorrent: "What struck me most"—Robert Graves is the witness—"was the number of dead horses and mules lying about; human corpses I was accustomed to, but it seemed wrong for animals to be dragged into the war like this."[31] One of the most horrifying of many horrifying passages in Remarque's *All Quiet on the Western Front* details the sharp, wounded cries and bellows of pain from injured horses stumbling erratically in no man's land—sounds that pierced the men with particular compassion, powerless as they were to go to the animals to despatch them mercifully.[32]

The warrior's compassion may be the spontaneous expression of young men in the midst of carnage or the more rule-governed response of the professional soldier who is scrupulous in adhering to wartime conventions. Evan Connell, for example, recalls messages exchanged between General George Armstrong Custer and a former West Point roommate, Thomas "Tex" Rosser, during the Civil War when they faced one another as opponents. Rosser's to Custer was addressed "Dear Fanny" (a nickname); Custer's to Ross, "Dear Friend."[33] Attempting to destroy one another on the field of battle, they remained good friends capable of a caring and humorous relationship to one another. It is difficult for anyone who has not participated in war to understand this *impersonal* attitude of respect that does not destroy affection, even between foes—and is difficult as well for some soldiers in war to understand. Often seen as the peculiar prerogative of the officer classes in modern European wars, the attitude I here describe is in fact more widespread. Sometimes it requires peace (or, better, the cessation of hostilities, *peace* being a problematic word as I take it up later) to flower fully.

In February 1985, 200 ex-Marines and 132 of their Japanese counterparts returned to Iwo Jima—on the fortieth anniversary of the dreadful prolonged battle for possession of the island—in a ceremony of remembrance and forgiveness. As former enemies embraced, one American said, "I can expel some of the demons that have haunted me for forty years. Now my memories are tempered. Now I can die in peace."[34] William Broyles, returning to Vietnam to confront some of his own ghosts, notes the "intimate combat" with the enemy, remarking, "I knew nothing about them." Finding the marker on a grave outside Hill 10, Broyles saw the name Ngo Ngoc Tuan, with the dates 1944–1969: "It was the grave of a man born the year I was, and killed the year I arrived in Vietnam. . . . And I stood, a tourist from the land of his old enemy, and looked upon his grave and thought how it might have been my own. . . . We had tried to kill each other, but we were brothers now." Broyles discovered he had "more in common with my old enemies than with anyone except the men who had fought at my side."[35] This sort of recognition is one shared by all individuals who have gone through a powerful experience together, one that punctures routine expectations, that distills experience through a luminescent rush of heightened emotion. Survivors of shared catastrophes speak of similar recognitions, as do women who fought in partisan and resistance movements. But the clearest and most consistent expression of human companionship amid destruction and fellowship that can transcend enmities is in the practice and literature of war, in stories of the sacrificial soldier and the professional brotherhood.*

---

* In this, soldiers are not that much different from other transnational identifications—scientist, for example, or doctor. Today there are international organizations of such epistemic communities that link up to promote cooperation on particular projects despite hostilities between their states. Similarly, feminists have argued that either *mothering* or simply *being a woman* suffices to constitute a transnational identity—sisters-not-in-arms.

The Literature of War

> It is true but of course women cannot suffer from it the way the men do, men after all are soldiers, and women are not, and love France as much as we do and we love France as much as the men do, but after all we are not soldiers and so we cannot feel a defeat the way they do, and besides in a defeat after a defeat women have more to do than men have they have more to occupy them that is natural enough in a defeat, and so they have less time to suffer.
>
> —GERTRUDE STEIN, *Wars I Have Seen*

> *Monday, August 3rd.* Today has been far too exciting to enable me to feel at all like sleep—in fact it is one of the most thrilling I have ever lived through, though without doubt there are many more to come. That which has been so long anticipated by some & scoffed at by others has come to pass at last—Armageddon in Europe!
>
> —VERA BRITTAIN, *Chronicle of Youth.*
> *The War Diary 1913–1917*

Vera Brittain's voice is lodged securely in the ranks of women pacifists and anti-militarists, and she has become a heroine to contemporary feminist anti-war activists and thinkers—not, however, without some ambiguity. In *Testament of Youth*, Brittain hankers after her wartime months in Malta—"during the War's worst period of miserable stagnation"—because Malta had come to seem a "shrine, the object of a pilgrimage, a fairy country which I knew that I must see again before I die. . . . Come back, magic days! I was sorrowful, anxious, frustrated, lonely—but yet how vividly alive!"[36]* But it is Brittain's wartime diaries for 1913 to 1917, published as *Chronicle of Youth*, with which I am here concerned. Brittain believed England must enter the war and could not remain neutral. Not to come to the aid of France would make England guilty of "the grossest treachery." Feeling much of the time as if she were dreaming, Brittain logged entries of how her father raved at her brother not to volunteer, while brother Edward, with a bit of class snobbism, told "Daddy" that, "not being a public school

---

* This expression by Brittain of some of the group ecstasy of wartime troubles analysts who see her covertly celebrating that which she enjoins. Brittain is simply being honest to the structure and texture of wartime experience.

211

man or having any training," he could not possibly understand how impossible it was for others to remain in inglorious inaction.[37]

Brittain understood the action required of war fighters in sacrificial terms. Disdaining her father's lack of courage in behalf of her brother, she noted the "agony of Belgium," remarking, bitterly, on *"the unmanliness of it, especially after we read in The Times of a mother who said to her hesitating son, 'My boy I don't want you to go but if I were you I should!'"* (italics added). Brittain's beloved, Roland, longed to take part in the war, finding in the possibility for sacrifice something "ennobling and very beautiful." After Roland died, Brittain capitalized the personal pronoun—"Him"—transfiguring Roland into a latter-day Christ in her entries about him. He died and became, in "His" sacrifice, a beatific figure for her: "On Sunday night at 11.0—the day of the month and hour of His death—I knelt before the window in my ward & prayed, not to God but to him. . . . Always at that hour I will turn to Him, just as the Mohammedans always turn to Mecca at sunrise."[38] Apotheosized in death, Roland lived on.

Wars are deeded to us as texts of a particular kind—a refrain we've heard before. When the texts are first-person accounts, real or imagined, they invite us to enter a war of words, to familiarize ourselves with the text and the texture of wartime experience. Novels numbered among the world's greatest—for example, Tolstoy's *War and Peace*—give presence, even nobility to the genre. War is an extraordinary laboratory; a high-pitched human experiment, dreadful and sublime, coarse and ennobling. No wonder we have so long been fascinated with it—*even* when the war story turns out to be one not of victory but of victimization.

Because women are *exterior* to war, men *interior*, men have long been the great war-story tellers, legitimated in that role because they have "been there" or because they have greater entrée into what it "must be like." The stories of women resistance fighters and soldiers have sometimes been told but have not attained the *literary* status of the great war novels by men—Stephen Crane, Erich Maria Remarque, A. P. Herbert, Ernest Hemingway, Norman Mailer, James Jones, Joseph Heller; nor have women written the most powerful war poems of the sort we associate with the names Rupert Brooke, Siegfried Sassoon, Wilfred Owen, or memorable autobiographical accounts on the order of Robert Graves, J. Glenn Gray, Ernst Jünger, Guy Sajer.[39]

Marguerite Duras's recently published memoirs join the highest rank for their sharp remorselessness and incisive capture of the way of being of the woman as resistance conspirator and home-front "waiter" for the return of her husband, a prisoner in a Nazi camp. The disintegrative fever of wartime is grasped by Duras in pellucid prose which takes the reader into the terrible darkness of the Second World War in Europe. The most poignant and generative recent account *of* war accounts is Paul Fussell's *The Great War and Modern Memory*.[40]*

Part of the female *absence* has to do with how war gets defined (where *is* the front?) and with who is authorized to *narrate*. Thus we find the first-person accounts by women who were "there"—in Vietnam—beginning to pour forth, in nothing like the numbers of male accounts but in greater profusion than previously because male fighters in that war have been represented, and have represented themselves, as *victims* above all—an identity to which women are interior. As well, because the "war at home" is not exterior to the Vietnam era but one of its central, defining features, women novelists such as Jayne Anne Phillips (*Machine Dreams*) and Bobbie Ann Mason (*In Country*), are telling the story of the war from the point of view of noncombatant female protagonists who were nevertheless "there" or were, in subtle but profound ways, victims of the war.[42] Isabel Colegate, in her 1981 novel, *The Shooting Party*, returned to the autumn of 1913, and depicted a prelude to the tragedy about to burst upon Europe in a story bubbling with premonitions in which the hunt—"male blood sport"—is emblematic for the "sacrificial note, the note of death, of blood" that the hunt itself signifies and presages more widely.[43]

Muriel Spark's seven-page short story "The First Year of My Life" is a condensed masterwork of ironic discourse, which features as its protagonist a baby "born on the first day of the second month of the last year of the First World War, a Friday." Born omniscient, as are "all of the young of the human species"—a power that gets "brain-

---

* Fussell has mused over why women haven't written more good war poems: bereaved women are, "next to the permanently wounded, the main victims in war, their dead men having been removed beyond suffering. Yet the elegies are written by men, the poems registering a love of soldiers are written by men, and it is not women who seem to be the custodians of the subtlest sort of antiwar irony. This seems odd, and it awaits interpretation."[41]

213

washed out of us" after the first year—Spark's baby doesn't smile as she watches "black-dressed people, females of the species, . . . saying they had lost their sons . . . careless women in black lost their husbands and brothers." On the baby's omniscient frequency, "the Western Front . . . was sheer blood, mud, dismembered bodies." Born into a bad moment, the baby continues to refuse to smile, puzzling those around her. Banging her spoon at figures of the dead and maimed, overhearing mourning of the deaths of war poets, finally, on the baby's first birthday, with the clamor, "Why doesn't she smile?" swelling to a crescendo, the baby smiles on hearing a report that Prime Minister Herbert Asquith in the House of Commons said the war had made all things new, a "great cleansing and purging it has been the privilege of our country to play her part. . . . That did it. I broke into a decided smile." Everyone around the baby concludes, "It was the candle on her cake." The baby knows better: "The cake be damned."[44] To really smile means to respond to the words uttered by Asquith—and, by extension, to those uttered by all who see a glorious new world as the outgrowth of massive carnage. The story is interesting, in part, because the ironic voice Spark has discovered, some sixty years after the event, is a voice largely foreign to women writers of explicit, historicized accounts of war, factive and fictive; Gertrude Stein is another exception.[45] Women *anti-war* writers sometimes deploy the ironic voice. Spark's story cuts several ways—against glorifiers of war *and* against utopian projections of peace.

THE FIRST WORLD WAR: WOMAN'S VOICE/MEN'S LIVES

When women have imagined war itself, however, it has frequently been in abstract, stereotypical tropes that bear little relation to war's realities. Fragments of these constructions are scattered throughout this text, from patriotic doggerel commemorating the death of Nathan Hale to First World War jingoist women poets imagining the Western front as a place of freedom even as men at the front found themselves "*im*mobilized . . . in trenches of death."[46]

Edith Wharton's account of *Fighting France*, published in 1915, and Gertrude Stein's *Wars I Have Seen* (1945) are literary constructions in which the lines between fact and fiction are purposefully blurred—at least Stein is knowing about what she is up to. All memoirs are, in Paul Fussell's words, "a kind of fiction, differing from the 'first novel'— only by a continuous implicit attestation of veracity or appeals to documented historical fact."[47] Looking around her, Wharton, who journeyed through France in 1915, sees buoyance and life, its "sudden flaming up . . . the abeyance of every small and mean preoccupation," clearing "the moral air as the streets had been cleared."[48]

Wharton's account is not, of course, exceptional. Men being sent off to the front, riding the crest of the jubilations of August 1914 in Europe, were lofted up on popular enthusiasms. But the actual experience of wartime soon crushed such sentiment, inviting bitter German soldiers (in Remarque's *All Quiet on the Western Front*) to curse angrily the teachers and elders who sent them off to carnage with abstract slogans and gestures; to envelop, in spite, their sanctioned designation, "We are the Iron Youth"—words that have become a bitter taste in their mouths, not an occasion for jubilant mobilization.

Such bitterness was foreign to Wharton's sensibility as she journeyed around France inscribing exaltation:

> *La France est une nation guerrière.* War is the greatest of paradoxes: the most senseless and disheartening of human retrogressions, and yet the stimulant of qualities of soul which, in every race, can seemingly find no means of renewal. . . . War has given beauty to faces that were interesting, humorous, acute, malicious, a hundred vivid and expressive things, but last and least of all beautiful. . . . War was the white glow of dedication . . . for the moment baser sentiments were silenced.[49]

The problem with this abstract construction is not that Wharton built it in the first place but that she had no way to take its measure against concrete experiences. She and the war passed one another by, as frequently happens for noncombatants who are not pressed upon as war fighters are. The war's worst hardships were yet to come; but those hardships, clashing with remembered imagination as a constructed narrative, both word and deed, are difficult to reassess from the lofty perch Wharton assumed in her celebration of the transformative power of war.*

---

* I don't mean to suggest that all male war memoirs and novels are up to the task of conveying

An instructive contrast is offered by comparing two novels of the First World War: Willa Cather's *One of Ours*, published in 1922, going on to win the 1923 Pulitzer Prize for fiction; and Ernest Hemingway's *A Farewell to Arms*, published in 1929. In Cather's novel, Claude Wheeler leaves a bleak existence at the core of which is a profound emptiness occasionally softened or penetrated by the love of his mother and the housemaid, Old Mahailey.[51] He gains brief intimations of another way of life when he spends a short time away at college and enters the homes of refined members of the middle class. He marries a temperance activist who shuns intimacy and takes off to be a missionary in China. In Claude's life on a farm in the American prairie land, nothing ever happens. Claude's future is indeed bleak; he would "become one of those dead people that moved about the streets of Frankfort." Things perk up as news of the war emanates to the American plains. Claude decides he must go "over to help fight the Germans," having been stirred up by stories of horrors perpetrated by the Hun and repeated as historic truth by Cather. The housemaid blesses his decision, "I knowed you would," she sobbed, "I always knowed you would, you nice boy, you! Old Mahail' knowed!"

His trip to Europe and through France is a voyage of discovery and familiarity. Although some of the flowers and the blueness of the sky remind him of home, life's adventure opens up for Claude, a young man who just two years earlier "had seemed a fellow for whom life was over." The war is a "golden chance." The French families who take in Claude and other American soldiers are patriotic, kind, welcoming, unresentful, offering clean beds and delicious food. ("To be so warm, so dry, so clean, so beloved!") The war took a little fellow and "gave him an air and a swagger." When Claude and other soldiers talk, it is always in well-rounded, complete sentences, peppered with exclamations and "Oh's." There are no sentence fragments, no indelicacies, no curse words, no epithets.

---

the reality they purport to depict. Falling victim to locutions and conventions one associates with a standard war story—one shorn of the irony that characterizes modern, disillusioned understanding—male war stories cast as hardbitten realism sometimes occupy an abstract distant perch, schematic and forced. Recent examples would include the series by W. E. B. Griffin, *Brotherhood of War*, with separate volumes on lieutenants, captains, majors, colonels, berets, and, finally, the generals; William Young Boyd's *The Gentle Infantryman*; and Scott C. S. Stone's *Pearl Harbor*, which begins: "December 6, 1941, was a lovely, calm day in the Hawaiian Islands. . . . It was a typical warm Saturday on a lush and slow paced island."[50]

Men "fall"—they are not killed—in Wharton's book. It is all very pleasant, antiseptic; wounds are clean; and the end is swift and merciful. The enemy is "the Hun," and several terrible stories are told about him. The war probably wouldn't make the world safe for democracy, but something important would come of it, some "new idea" would come "into the world . . . something Olympian." Claude falls, dying instantly. His body is retrieved and sent home. His mother reassures herself by reading his letters: "For him the call was clear, the cause glorious." He died holding "beautiful beliefs," and it is as well he did and was spared the dread of disappointment. He no doubt hoped and believed too much but, by the grace of God, was saved from "some horrible end," a fate worse than death—disillusionment. The price, for Cather's Claude, is right.

What is striking about Cather's account is not its enthusiasm for the war, for that was widely shared, but its literary reinscription of that jubilant innocence as the war dragged to its bitter end, plummeting from all the high hopes that had marked its beginning, illusions from which Europeans had become rapidly disenchanted—or at least those fighting the war had. Wholly without irony, Cather's prose grates. Its apotheosis of war is sicklied with an abstract sentimentalism. She was not, of course, wrong to see war as a great transformative experience. But Claude goes from being one sort of wimp, to use modern parlance, to being another. Claude is a Beautiful Soul.

The year Cather's novel hit the bookstores, a dispatch by Ernest Hemingway appeared in the *Toronto Star Weekly* for 8 December entitled "War Medals for Sale": "What is the market price of valor?" a man asked. The pawnshop owner told him that men came in every day to sell medals, but "we don't buy medals from this war."[52] A short story of Hemingway's from the 1920s, "Soldier's Home," tells the sorry saga of young Krebs, from Oklahoma, who enlisted in the Marines in 1917 and served for two years. He returns to his home feeling the "need to talk but no one wanted to hear about it. His town had heard too many atrocity stories to be thrilled by actualities. Krebs found that to be listened to at all he had to lie. . . . A distaste for everything that had happened to him in the war set in because of the lies he had told."[53]

"If the war didn't happen to kill you," George Orwell once remarked, "it was bound to set you thinking." Set to thinking, and writing, Hemingway produced *A Farewell to Arms*. His protagonist, Lieutenant Frederic Henry, comes to hate *the* War, that grandly constructed event so much at odds with the bitter and the exhilarating experiences he has had. Characters throughout the book tell of war hatred and confusion about why the war goes on. His protagonist quits the war ("I had made a separate peace") and, determining to forget the war, rows with his pregnant lover, the nurse Catherine Barkley, to the safety of neutral Switzerland. Unlike Cather, who unselfconsciously uses words of sacrifice, honor, ecstasy, freedom, Hemingway enunciates what became the modernist understanding—bitter, ironic:

> I was always embarrassed by the words sacred, glorious, and sacrifice and the expression in vain. We had heard them, sometimes standing in the rain almost out of earshot, . . . and had read them. . . . There were many words that you could not stand to hear and finally only the names of places had dignity. . . . Abstract words such as glory, honor, courage, or hallow were obscene beside the concrete names of villages, the numbers of roads, the names of rivers, the numbers of regiments and the dates.[54]

To the irony of which Hemingway is here master must be added an additional ironic note: that is, in the matter of life giving, life taking, and history's gender gap, male modernists offer the critical distancing from war and the reflective puncturing of war myths that most powerfully served to defeat the simplistic, hollow heroics characterizing the Western world's plunge into the First World War. Women who wrote of war hadn't (for the most part) had the experiences that made the men wary and suspicious. But here, too, the gap is narrowing—or has narrowed—for Americans whose most distressing emblem of war is Vietnam. Vietnam is even now in the process of being reconstructed as a story of universal victimization—of Vietnamese by us; of our soldiers by the war—and by us when we didn't welcome them home; of our nation by the war at home and *the* war; of wives and girlfriends by disturbed veterans; of nurses by the war and later nonrecognition of *their* victimization. Although the experience of the male vet is preeminent in this cultural reinterpretation, women are less out of it than in previous shared understandings.

Men: The Militant Many/The Pacific Few

VIETNAM: THE VETERANS SPEAK

> It's like when you're walking down a dark street at night, and out of the corner of your eye you see somebody getting hurt in a dark alley. But you keep walking on because you think it don't have nothin' to do with you and you just want to get home. Vietnam turned this whole country into that dark street, and unless we can walk down those dark alleys and look into the eyes of those men and women, we're never gonna get home.
>
> —BRUCE SPRINGSTEEN, *Boston Globe*

> I gave my dead dick for John Wayne.
>
> —RON KOVIC, *Born on the Fourth of July*

We—as a people—listen now as the Vietnam vets speak to and from their anguish. Their wartime deeds having been recast as an experience of victimization, in turn psychologized and translated into the therapeutic language of post-traumatic stress syndromes and repression and healing, they are set apart even from other veterans. Middle-class Americans valorize these self-understandings if they take a form that we can name "anti-war."* At the moment, we cannot seem to get enough of the tales of suffering and sacrifice told by vets who are haunted, some apparently doomed, others rising to the surface.[56] Is it possible we envy them their suffering and their sacrifice? Where does our *need to know* come from? It seems that what *they* were doing makes for a gripping tale after all. They went to the edge and fell over. We think: the risks were *real*, the stakes *high*, the deeds *terrible*, the drama stark, the suffering *unbelievable*, the horror *incomprehensible*—and that is the fascination. For a time it appeared that Vietnam might go unnamed, like female violence, living in a semiotic twilight zone save for abstract defenses of "our involvement" or denunciations of the same thing. But that isn't the way it has turned out. We've devised a way to appropriate it: Vietnam has become an "experience," unique in the annals of American war narratives.[57]

The stories being told are not classic war stories of triumph and honor. They are, instead, equally riveting tales of violence that crosses

* Of one thousand two hundred in the Harvard class of 1970, fifty-six entered the military and two went to Vietnam. In 1971, a majority agreed with the Harris poll that men who went to Vietnam were "suckers, having to risk their lives in the wrong war, in the wrong place, at the wrong time."[55]

over the boundary separating rule-governed fighting from "free fire zones," cohabiting with myriad debasements—cutting off enemy ears as war trophies, mutilating or marking corpses, wasting prisoners, burning villages, assaulting women—then emerging to tell what it was "really like." For all the horrors these soldiers committed, they met their full measure in turn. The war transformed them. When they departed "in country," it was as "changed" men and women who, in a few "extreme cases felt that the experience there had been a glorious one, while most of us felt that it had been merely wonderful."[58]

What I am trying to get at here is the way we serve as an audience for stories of the Vietnam experience.* It is a matter not so much of extending our empathy (I agree with Michael Herr that "those people who used to say that they only wept for the Vietnamese never really wept for anyone at all if they couldn't squeeze out at least one for these men and boys when they died or had their lives cracked open for them"[59]) but of our need to appropriate *their* experience, to draw it within the familiar circle of our understanding. Not surprisingly, given the themes that dominate contemporary American culture, the Vietnam vets' struggle for self-definition emerges as a form of individual and collective *therapy*, a public and private discourse. Their needs can be structured to speak publicly to those whose lives were not cracked open violently, helping them to approach those who have been internal "others"—not just most women but the majority of males of their age group. Vietnam war fighters and home-front fighters *of* the war were young men, and women, of the same generation separated by class privilege and disparate self-definitions, estranged from one another by conflicting notions of duty, loyalty, patriotism, and purpose, and further alienated by war. Now they (the Vets) seem to be allies in the fight against war. As a society, we have accepted them as wounded guides to a *terra incognita* which we declare loathsome but *find* compelling—though that is best not talked about, like female lynch mobs.

Whether the current accessibility of the "Vietnam experience" in and through the language of victimization, estrangement, therapy, and healing will, over time, narrow the gender gap remains to be seen. But

* I count one hundred newspaper clippings on the table beside me at the moment, dating from 1981 to the present. In the Vietnam era, 2,900,000 served; 300,000 were wounded; 55,000-plus were killed; their average age was nineteen.

the irony of authorization (*Who* gets to speak? *Who* listens?) remains. By all accounts the most effective voice against the Vietnam War were those of the vets themselves during the Winter Soldier hearings in 1970, when they threw away their medals, casting them over the fence around the White House: "We now strip ourselves of these medals of courage and heroism. We cast them away as a symbol of dishonor, shame and inhumanity."[60]* (See photo section, #12.) Because these were men who had fought, they were granted an immediate authority others lacked. It is not all that different from the Mothers' Strike for Peace in the 1960s, when women, pressing their case against atmospheric nuclear testing, chose a most provocative symbol to combat the pollution of that most hallowed of all liquid substances—mother's milk. A skull-and-crossbones emblem on a baby's bottle is as powerful as a veteran in his wheelchair throwing away the insignia of war.

Structures of Experience:
The Good Soldier/The Good Mother

While the bombardment was knocking the trench to pieces at Fossalta, he lay very flat and sweated and prayed oh jesus christ get me out of here. Dear jesus please get me out. Christ please please please christ. If you'll only keep me from getting killed I'll do anything you say. I believe in you and I'll tell every one in the world that you are the only one that matters. Please please dear jesus. The shelling moved further up the line. We went to work on the trench and in the morning the sun came up and the day was hot and muggy and cheerful and quiet. The next night back at Mestre he did not tell the girl he went upstairs with at the Villa Rossa about Jesus. And he never told anybody.
                                        —ERNEST HEMINGWAY, "Soldier's Home"

Too many people learn about war with no inconvenience to themselves. They read about Verdun or Stalingrad without comprehension, sitting in a comfortable armchair, with their feet beside the fire, preparing to go about their business the next day, as usual. One should really read such accounts under compulsion, in discomfort, considering oneself fortunate

---

* In the Winter Soldier Investigation, convened in Detroit in February 1971, veterans testified and confessed publicly to crimes they had committed in the line of duty.

not to be describing the events in a letter home, writing from a hole in
the mud. One should read about war in the worst circumstances, when
everything is going badly, remembering that the torments of peace are
trivial, and not worth any white hairs.

—GUY SAJER, *The Forgotten Soldier*

Hemingway's terrified combatant in "Soldier's Home" pleads ("jesus
please get me out") not unlike women in labor who often similarly
plead ("someone help me, oh stop the pain, please help").[61] The shelling
over, the baby born, the terror is quickly shoved into a dark background
as if it has never happened. Men find at odds the language available to
express their terror and the terror itself; women fall silent about the
pain of childbirth—at least much of the time. The silence is culturally
sanctioned, or has been, rather like a well-guarded but widely shared
secret.

It is my hunch that the experienced inadequacies of language and
subtle interdictions on speaking of certain things—unless one lies, like
Krebs in Hemingway's short story—are likely outcomes of powerful
boundary events, a particular structure of experience discontinuous
with the expectations of everyday life. Here I take my cue from Eric
Leed's *No Man's Land: Combat and Identity in World War I*, a path-
breaking discussion that offers insight where soggy confusion is often
the rule.[62] Some expectations about soldiering and mothering are shared
and out in the (cultural) open, so to speak: the soldier is expected to
sacrifice for his country as mothers are expected to sacrifice for their
children. This isn't quite a sacrificial symmetry, to be sure. Most women
do not forfeit life itself but they forfeit a version of what their lives
might have been, as do male combatants who are forever changed by
what they have been through.

Uniting the two experiences is *duty* and *guilt*. The soldier and the
mother do their duty, and both are racked by guilt at not having done
it right or at having done wrong as they did what they thought was
right. The warrior's deep ache of guilt is also a mother's keen burden
and joy. One might have acted differently and a buddy been saved.
One might have lived up to this ideal and a child spared that trauma
or this distress. Such guilt is not, J. Glenn Gray argues, a personal
psychological burden so much as an acute awareness of human freedom
and a failure to live up to one's ideal of the good.[63]

Also shared is a slippery slide toward forgetting, on one end, toward remembering in nostalgic and sentimental ways, on the other. What is there about the experiences of childbearing and mothering, of arms-bearing and soldiering, that invites both sorts of reaction? Both are boundary experiences that forever alter the identities of those to whom they happen, or through whom they take place. War experience, writes Leed, is "an experience of radical discontinuity on every level of consciousness."[64] It is not a place one has been before. The worlds of "peace" and "war," and the worlds of "childbearing" and being "childless" (or "free") in current parlance, are, on some profound level, incommensurable. You have to *experience* it to know it. Just as romantic visions of what "war was and what it would mean" seem never to go away, so similarly romantic images of mothering grip our collective imagination despite repeated accounts that detail the gritty reality.

Leed writes of "the liminality of war"; war is a rite of passage, a boundary experience, profoundly reshaping its veterans.[65] War experience is concrete, dirty, filled with pollution and fluids, loss of control over one's body, experiences of keen vulnerability and absence of privacy. "The soldier is on friendlier terms than other men with his stomach and intestines," writes Remarque. "Three quarters of his vocabulary is derived from these regions. . . . It is impossible to express oneself in any other way so clearly and pithily."[66] Soldiers, like mothers, are involved with food, shit, and dirt. He throws himself on "Earth!—Earth!—Earth!," his "only friend," his shelter when he is under attack. He concentrates obsessively on his stomach and bodily "intactness."[67]

Mothers, and children, are similarly immersed in worlds revolving around stomachs, bodily harm or well-being, the search for protection. Physicality defines the lives of mothers and soldiers, but expressions of this physical concreteness often shocks or dismays others: it isn't the way we want to think about things. Each is attuned to particular sounds—the whimper, the cry, the scream, the whine, the hiss, the explosion—and subject to both empowerment and astonishment at being able to respond appropriately, to come through safely ("I had no idea. . . ." "I was not prepared for . . .").

The ideal of the Good Officer like that of the Good Mother emphasizes *caritas*, Christian caring, giving such action an "almost religious" aura. The abstract words—like *honor, glory, mother love*—give way, when the soldier or the mother is *in the throes*, to the concrete activity of staying alive, or keeping others alive, of tending to *this* moment, *this* comrade, *this* child—no abstraction but a living being to whom I am responsible. *Interior* to this world, being *in* the experience and not in some abstract evocation of it, soldiers and mothers frequently talk about war and mothering as a force beyond or greater than themselves, an event or a structure of experience and events "that has no author"; one is instead "enmeshed in it, the event had become a 'text,' the correct reading of which was a matter of life and death." War and mothering have been worlds that *enfold* men and women, worlds with their "own logic, connections and incongruencies."[68]

Some might cavil at this constructed similarity of structured experience, arguing that the soldier's injunction to kill is a terrible mirror image of the mother's imperative to preserve. But even here things are not crystal clear. Men's experience of war is defensive, a story of aggression held (for the most part) in check, a tale of trying to protect, to save, to prevent. The reality of modern combat structures war as a highly constrictive activity in which defense rules: "The realities of war forced a curtailment of aggression, a ritualization of violence, and the holding back of hostile impulses"—not their unleashing.* The warrior who emerged out of the First World War was not an "offensive" but a "defensive" personality, "molded by identifications with the victims of a war dominated by 'impersonal' aggressors of chemicals and steel."[70] The war lover on a killing binge was someone who had "lost it" just as the defensive mother who batters her child has lost it, having gone from protector to attacker.†

The defensive warrior, trying to save and preserve, if necessary sacrifice himself, rather than kill—kill—kill, is sickened by the ferocity of bellicose noncombatants—just as mothers are put off by professional advice on how they are to do their job, on what is important, and on what is *really* going on with their children if this advice is couched by a distant professional who "hasn't been there." Just as soldiers fre-

---

* According to Leed, "at least one-half of tactical thinking, and more than one-half of military activity, is occupied in frustrating the aggression of the enemy."[69]

† During the First World War, what had been predominantly "a disease of women before the war—neurasthenia—became a disease of men in combat . . . physicians immediately recognized the kinship."[71] *Neurasthenia* was the name for a generalized anxiety syndrome, stemming in part, so it was alleged, from enforced *inaction*.

quently cannot communicate what "being there" is all about (Robert Graves found home intolerable because there was no common ground between himself and others to talk about the war; when asked by his mother, "Was it very bad out there, Paul?" Baumer, Remarque's protagonist, lies, "No Mother, not so very. There are always a lot of us together so it isn't so bad"[72]), mothers find themselves similarly frustrated in their attempt to express the burdens and joys of mothering to others (most often men). Women are excluded from war talk; men, from baby talk. Men conceive of war as a freedom "from" and find themselves pinned down, constrained; women see mothering as the ticket to adulthood and find themselves enmeshed in a dense fabric of responsibility that constrains even as it enables.

Perhaps we are not strangers to one another after all.

7

Neither Warriors nor Victims: Men, Women, and Civic Life

I went out, peace seemed imminent. I hurried back home, pursued by peace. It has suddenly struck me that there might be a future, that a foreign land was going to emerge out of this chaos where no one would wait any more.... Everyone's impatient because peace is so slow in coming.... The whole world is waiting.

—MARGUERITE DURAS, *War*

THE LOCUTION *warriors/victims* comes from remembering Albert Camus's writings and his insistence that as he would not be a victim, so he would not be an executioner; from recalling his warning that today's victims can turn and have turned into tomorrow's executioners—an all too likely temptation.[1] The locution also derives from Freeman Dyson's discussion in *Weapons and Hope* concerning the present incommensurability of two worlds he tags the worlds of the warriors and of the victims, respectively, each giving rise to its own discursive style. The world of the warriors, he notes, promotes a style that is "deliberately cool, attempting to exclude overt emotion and rhetoric ... emphasizing technical accuracy and objectivity." The world of the victims is not male dominated, as is the warriors' domain, but is "women-and-children-dominated.... The warriors' world describes the outcome of war in the language of exchange ratios and cost effectiveness; the victims' world describes it in the language of comedy and tragedy."[2] Mostly tragedy. Frequently laced with eschatological fears. More and more victims peer over the edge of the abyss and come back convinced that an apocalypse looms. Soon.

This chapter, however, is based on the premise that we will as a country, that we will as an international community, be around for a while, neither disintegrating to match the worst fears of many nor rising to meet the high hopes of others. This chapter, in other words, is about politics. Politics is the work of citizens, human beings in their civic capacities. In such capacities, men and women over the years have struggled to eliminate warfare and to attain a hope of universal peace. The efforts, cast as absolute goals, have failed. I begin with the story of one such attempt—the liberal dream of a rational, peaceful international society—and go on to assay feminism's historic war with war. There follows speculation concerning modes of discourse that might help sustain hope and underscore a civic ethos to combat our current deadlock and provide for a freer play of individual and civic virtues than we now enjoy. I meditate on the "problem with peace" and love of country, on civic and personal identity, as I reach the end of these reflections.

The Liberal Conscience

> There is a morning of reason rising upon man on the subject of government, that has not appeared before. As the barbarism of the present old governments expires, the moral conditions of nations with respect to each other will be changed. Man will not be brought up with the savage idea of considering his species his enemy.
>
> —THOMAS PAINE, *Rights of Man*

In the seventeenth and eighteenth centuries, concern increased over the incessant state of war in which European nations were embroiled. The rise of liberal humanism, indebted to the writings of Erasmus, Thomas More, and several Renaissance humanists, increasingly attributed war to the excesses of rulers or to atavistic holdovers from feudal warrioring classes and norms. The international lawyers of this period worked to codify rules of war in a way aimed at civilizing war if not eliminating it. The belief that wars arise because of *misunderstanding* took hold in this era and, with it, the concomitant hope that if people just communicate they will be able to adjudicate their differences amicably.

Another conviction that helped to construct what Michael Howard calls the "liberal conscience" is the insistence that the growth of international commerce, the rising *interdependence* of nations, would trail peace in its wake as a kind of side benefit.[3] Montesquieu was firm in his belief that peace is a "natural effect of commerce" as nations that trade together become "mutually dependent."[4] A similar brief was lodged by Joseph Schumpeter whose influential writings made the case for capitalism on the basis of its beneficial consequences, commercial societies being devoid of a warlike spirit save as a lingering holdover from feudal, precapitalist mentalities and social formations.[5] As Albert Hirschman argues, the interests were to overcome the passions, especially the passion for conquest."[6]

John Stuart Mill, in this as in other matters, put great faith in the triumph of Reason and was, consequently, rather buoyant from his vantage point in the nineteenth century. Writing in 1848, Mill opines:

It is commerce which is rapidly rendering war obsolete, by strengthening and multiplying the personal interests which act in opposition to it. And it may be said without exaggeration that the great extent and rapid increase of international trade, in being the principal guarantee of the peace of the world, is the great permanent security for the uninterrupted progress of the ideas, the institutions, and the character of the human race.[7]*

Other twists and turns in liberal hopes included, and include to the present moment, calls for national freedom as a right of peoples and a way to ease tensions; for education to peace, weaning people, especially would-be leaders, away from warlike enthusiasms and yearnings;

* Immanuel Kant in 1795 also believed peace would come about given the spread of republican governments and the growth of commerce and trade. I take up his plan for "perpetual peace" in a later section.

and for organizing the enlightened classes of all nations into a sure and secure force for peace on a transnational basis.*

War was deemed *irrational*. It followed that wars were outmoded and should be superseded by a more sensible way of doing things. The high-water mark of this effort—though we may be in another now—was reached just before the debacle of the Great War. Prior to 1914, some "194 treaties containing provision for arbitration were signed. . . . There were 425 Peace Organizations in being throughout the world. . . . University Peace Congresses were held annually from 1892"; and, in addition, innumerable "functional links" had been established in "the fields of trade, travel and communications." The Carnegie Endowment for International Peace was established in 1910 with a fund of ten million dollars to study the causes of war, to promote international law, and to influence public sentiment against war; it persists despite intervening cataclysms.[8] So, too, do various women's peace efforts which I will look at in a moment.

Liberal hopes spring eternal, resounding in activities embodied in the following headlines gleaned from recent local and national newspapers: "Groups Protest Toys with War Theme"; "International Group of Physicians Gathers in Virginia to Warn of Nuclear War's Effects"; "The Sant Sani School Blends Love of Learning with Philosophy of Human Respect"; and from journals like *International Journal on World Peace*, which features articles on how to foster and structure a just peace through international law and organization and by convincing people once and for all that war is barbarian and irrational and should be abolished.

Liberals can point to the "separate peace" that exists between and among liberal states—the United States, Canada, Great Britain, Sweden, Norway, Australia, New Zealand, France, the Netherlands, and others—as support for the contention that liberal states are more peaceful—at least in their dealings with other liberal states. What Schumpeter called "bloody primitivism" seems not only ongoingly to have its due but to be given many a day—to the consternation of liberal-rationalists. Yet liberal hopes—which incorporate at their heart a conviction that national estrangements can be overcome, that a generic humanity might

---

* Of course, such groups are not limited *de jure* to the educated middle classes, but *de facto* that is always the way it has worked out.

one day recognize where its true interests lie—die hard. These hopes have animated feminism from its inception, for feminism, together with the liberal conscience here sketched, arose in tandem out of the intellectual and cultural movements of seventeenth- and eighteenth-century Europe and America. But before I turn to feminism, one resonant exemplar of the liberal conscience deserves detailed attention.

In 1910, William James wrote a leaflet (no. 27) for the Association for International Conciliation, one of the dozens of pre-war peace movements I have noted, in which he called for the "moral equivalent of war."[9] James begins crisply: The war against war "is going to be no holiday excursion or camping party." Why? Because "military feelings are too deeply grounded" to give way without a long struggle. War is life *in extremis*, and history "is a bath of blood." James depicts Greek history as a "panorama of jingoism and imperialism." Pugnacity has been "bred . . . into our bone and marrow." For all the liberal truisms he repeats, James is on to something important. He understands that war can present itself as a "rescue from the only alternative supposed . . . a world of clerks and teachers, . . . of 'consumer's leagues' and 'associated charities,' of industrialism unlimited and feminism unabashed." But history would be insipid if there were no prizes to the daring; no risks for the audacious. We are afraid to let go of war possibility, of the war regime, because we fear even more the prospect of a sterile peace. There is, James argues, but one alternative—a *moral equivalent* of war—in recognition of the widespread human taste for adventure, for sterner stuff at odds with the "mawkish and dishwatery" tastes of "present-day utopian literature." James would breed, through a nonviolent war—a war, in his vision, against *Nature*—the discipline and collective civic unity thus far only war has spawned. (His war against *Nature* involved conscripting youth to work in mines; build tunnels, railroads, and factories; even do housework; for James this was part of "the immemorial human warfare" against *Nature*.)

Although the prospect of a collective war against *Nature* seems nearly as naïve—or even unnecessary, for we have done quite well in that regard without mobilizing armies of youths for the purpose—as the watery utopianism many peace people advocated, James hit upon a conundrum that haunts peace movements and peace rhetoric: How do you garner people's energies in grand efforts, offering moments of collective epiphany thus far attained, unevenly to be sure but unequivocally, in wartime? Given that the binary opposition between war and peace is central to the liberal, bourgeois world view, the question *must* be as James so eloquently couched it. For what one seeks is wartime enthusiasms without war. What one strives for is war's generative powers without its destructiveness. Whether this is a dream worth dreaming I will take up as this discussion moves along, beginning with a lengthy look at feminism and war.

Uncertain Trumpet: Feminism's War with War

Over the past year, in one poll after another, women have staked out a clear position—against Reaganomics, against nuclear arms—and gradually men have drifted over to share those beliefs. . . . For whatever reasons—because of our culture, because of our history or because of motherhood—nonviolent convictions are more pervasive among women.
—ELLEN GOODMAN, *Washington Post*, 22 May 1982

It seems to me that one of the risks of the much-heralded peace gap is that women may rest on their moral superiority. It is easy, after all, to stake the high ground—peace—when you aren't slogging through the daily mud of foreign policy. . . . Women have had a certain luxury in being the outsiders. It's the luxury of not being directly involved, not being responsible.
—ELLEN GOODMAN, *Washington Post*, 1 November 1983

The subtle shift over a year's span in the position of a liberal feminist, the syndicated columnist Ellen Goodman, helps illustrate a deep feminist dilemma. From its inception, feminism has not quite known whether to fight men or to join them; whether to lament sex differences and deny their importance or to acknowledge and even valorize such differences; whether to condemn all wars outright or to extol women's contributions to war efforts. At times, feminists have done all of these things, with scant regard for consistency. Feminism moves along a number of planes: as the action of women in and on the world; as abstract theories and utopian evocations; and as a story of self-conscious feminists breaking down extant barriers to take their place in previously

231

all-male institutions—for example, the military. To be sure, the vast majority of women in the armed forces of the United States, at present, do not identify themselves as feminist; but because their identities and actions have been given a feminist gloss and representation, it seems appropriate to take up military women under this general rubric. The story of feminism and male/female identifications is, as I have previously suggested, a tumult of ongoing encounters with a long, grand genealogy—from the prototypical maternal figure, the Madonna, to the exemplary woman warrior, Joan. As discourse, feminism is not just a series of explicit endorsements but a cluster of implicit presumptions guiding rhetorical choices and controlling dominant tropes and metaphors.

The reflections to follow are being written at a time of flux when several ways to occupy the category *woman* present themselves, though only a few get the stamp of approval from one variety of feminism or another: "right to fight" feminists, who are endorsed by integrationist feminism incarnated by the National Organization for Women; revolutionary women in Third World countries, approved by most mainstream feminists and feminist anti-militarist groups save for absolute pacifists; the woman peace activist, sanctified by a plethora of old (Women's International League for Peace and Freedom) and new ("Greenham Common," as a generic for all-women peace encampments and direct action) efforts. At the anti-war end of the pole, pressure is put on military women to decamp; at the liberal-egalitarian end, urgent compulsions are reversed as women who overidentify (on this reading) their femaleness with peacefulness find this identification challenged or at least fretted about as simply another way to reinforce old stereotypes. All forms of feminism put pressure on "sleeping with soldiers," on the age-old union of Eros with Mars that makes of men in uniform attractive targets for libidinal fixation.*

A polyphonic chorus of female voices whose disparate melodies are discernible sounds now in the land. Among the many voices are latter-day Antigones ("Hell, no, I won't let *him* go"); traditional women ("I don't want to be unprotected and men are equipped to do the protection"); the home-front bellicist ("Go, man, go and die for our country");

---

* The evidence on this score is so pervasive it surely needs no rehearsing again here. Women have written and spoken of it, and so have men—repeatedly.

the civicly incapacitated ("I don't rightly know"); women warriors ("I'm prepared to fight, I'd like to kick a little ass"); and women peacemakers ("Peace is a woman's way"). Each of these voices can be construed as the tip of a pyramid descending on either side to congeal into recognizable social identities that sometimes manifest themselves as movements: a feminist revolution to end all wars (the penultimate earthly triumph of the victims); a socialist revolution to end all wars (the earthly triumph of total justice); a psychological revolution (the "hearts and minds" approach), with men being reconstructed along pacific lines; or, more modestly, a feminist *movement*, short of revolution, to gain female parity of power/force with men. Then, of course, there are women fighting changes, seeking, instead, a resurrection of traditional identities and a restoration of old complementarities. At the moment, feminists are not only at war with war but with one another, as well as being locked in combat with women not self-identified as feminist.

HISTORIC PEACE CAMPAIGNS

Women have been linked to peace campaigns at least since the Middle Ages, to anti-war sentiment since the Greek tragedies. But thus far women have been unable—whether as liberal humanists, socialist internationalists, or feminists—to forstall state conflicts. Not because of any lack of effort. An irony in women's peace campaigns, or one of the ironies, is this: wars that engender the creation of a plethora of women's organizations I noted at various points in preceding chapters also give rise to women's peace efforts. For example: swamped by the enormity of the First World War and the energies it canalized, the story of the Women's Peace Party, forerunner to the Women's International League for Peace and Freedom and canonized grandmother to the Greenham Common Women's Peace Camps, is less well known than it ought to be. The divide between peace women and war women in England made sharply visible a pre-war cleavage between the militant suffragettes of the Women's Social and Political Union, who believed in the occasional use of violent methods to gain their ends, and the suffragists, who were opposed to all forms of violence, including violence against property. With the war, the suffragettes tended toward jingoism; the suffragists were more likely to become peace campaigners.[10] In the United States, the Women's Peace Party was linked to, but not wholly subsumed under, the suffrage effort, although item 6 on the party platform called for the "further humanizing of governments by the extension of the suffrage to women."[11]

My intention here is not to recall a history well documented by others but to make explicit the markers of social identity and structures of discourse embedded within and giving rise to such efforts. At the height of the Women's Peace Party movement in the United States, there were some 165 group memberships totaling 40,000 women. The American party was one section of the Women's International Committee for Permanent Peace, which had branches in fifteen countries. Resting securely on the conviction—in Jane Addams's words—that feminism and militarism are in eternal opposition, the notion that women, as mothers, have a vested interest in peace, and the corollary insistence that women must assert until they triumph the "ultimate supremacy of moral agencies," were widely pronounced.[12]

Anti-war feminists pushed for continuous arbitration, for a negotiated peace short of total victory, and for a peace settlement shorn of vindictiveness. In this, and other ways, feminists in this period located themselves *within* the larger frame of pacifist/just-war discourse. Women peace campaigners promulgated internationalism, as a worldwide concatenation of peace-loving peoples, especially women, to bring into being the conditions for permanent peace. After the war, women were influential in pressing for the Kellogg-Briand Pact (1928) declaring war illegal and were on the National Committee on the Cause and Cure of War, which collected ten million signatures on a disarmament petition in 1932. Contemporary women's peace efforts look back to these, and other instances, of female activism as historic warrant for their own efforts. In the past six years, women-only peace camps have been created. The identities they celebrate, the oppositions they lodge, go back to these earlier insistencies as action and as discourse—as a brief look at the lives of two very different anti-militarist women of this era helps make clear.

WOMAN AS PEACE ACTIVIST, WOMAN AS SECRET OUTSIDER: JANE ADDAMS AND VIRGINIA WOOLF

Jane Addams and Virginia Woolf shared certain middle-class convictions, including a great trust in the power of moral agency to transform consciousness and, by extension, social life. In her classic tract *Three Guineas* (1938), Woolf held that men and women are different.[13] Men are drawn to bellicosity; women, to a more pacific stance. Thus it has always been; thus, presumably, it will be until male belligerence is disarmed as women cease to pay it any mind by practicing studied indifference toward dominant males. Addams, although a more complex political thinker, repeated one of bourgeois society's structuring cleavages—between war and peace—recasting it, as I already noted, as the eternal opposition between feminism and militarism. Because today's feminist peace activists are deeply indebted to earlier modes of action and thought, it is worth pausing for a moment to muse on the reflections of Addams and Woolf.

Jane Addams frequently turned her attention to the problem of women and war. In *The Long Road of Woman's Memory* (1916), *Peace and Bread in Time of War* (1922), and *Newer Ideals of Peace* (1907), she tied women's pacific sensibilities to women's nurturing functions which give them capacities and sensibilities desperately needed by the wider society.[14] Addams wrote movingly of the "maternal affection and solicitude in woman's remembering heart" and insisted that women have an "imperative to preserve human life," which she traced back to the evolutionary beginnings of human society and women's role in creating the social forms of *settled* existence.[15] Addams's is a sophisticated statement of women as pacific Other, a variant on the Beautiful Soul, which she attempts to ground historically and anthropologically and to bring to bear on public discourse.

But the idealizations she embraced weakened the cogency of her case by presuming, for example, that the extension of suffrage to women would automatically humanize governments. In asserting the "eternal opposition" between feminism and militarism, Addams constructs feminism as a collective Other, countering the collective embodiment of the warrior mentality and a politics of force and coercion. Implicitly

235

a bargain has been struck that gives her a license to evade many tough questions about power and responses to the use of force. But Addams at least keeps women in contention in the civic arena, fighting as *public citizens* for their ideals.

Virginia Woolf, on the other hand, valorizes women as outsiders, embracing a vision of female *withdrawal* into a fantasized society of the indifferent. Woolf's ex-patriotic plea—reinscribed by contemporary feminists for her resounding claim, "As a woman, I have no country. As a woman I want no country. As a woman my country is the whole world"—would free women from all "unreal" loyalties, including national loyalty, by encouraging them to "experiment not with public means in public but with private means in private."[16] The feelings that energize her tract are those of betrayal and resentment. Her solution is a vision of secret Lysistratas who have made a separate peace. Filled with clichés some feminists have criticized as the ideology of True Womanhood, Woolf ironically celebrates much of that vision, with this exception: she would spring women from their middle-class homes to enter professions one by one where, by living up to their luminescent code, they will transform society over time in a peaceful, noncompetitive direction.

While Woolf has previously killed the being she calls the "angel in the house," that angel has not, in fact, been destroyed—but has simply been relocated, as the daughters of educated men march into society. The double standard is celebrated: middle class women *are*, in fact, morally superior. They are to retain their purity as they pursue their individual ends in the professions, finally bringing their imprint to bear upon the world entire. Although important feminist critics, including Elaine Showalter, have criticized the language of *Three Guineas* as "empty sloganeering and cliché" and as a tract guaranteed to keep things as they are, it retains its vibrant power to incite contemporary feminists to action based upon a sharp cleavage between male/female psyches, consciousness, politics, morality, and purpose. The difference is that today's feminist/women peace activists organize and agitate collectively, rather than chasing the chimera of Woolf's secret outsider; they would realize rather than derealize the citizen.[17]

Neither Warriors nor Victims: Men, Women, and Civic Life

FEMINISM AS MOBILIZATION: STATES OF WAR,
PROMISES OF PEACE

It would be impossible to survey the whole of contemporary feminist discourse in a few pages; however, it is possible to *situate* feminist arguments on questions of war and peace inside one of several frames. Interestingly, feminism reproduces many assumptions that structure the discourses of realism and just war, respectively, re-creating prototypical characters and figurations. Thus, contemporary feminism has both its "Machiavellian moment" and its Beautiful Soul reiterations. A characteristic of much, though by no means all, current arguments is their totalizing nature: women must wage a *total* war against *every aspect* of patriarchal society; or women must struggle for *world* peace, for *total* disarmament. Women are mobilized by these diverse interventions, locating themselves *inside* a universe of collective self-definitions. Apocalyptic visions are pronounced by some; incremental reforms endorsed by others.

The internal structure of inherited war discourse and mobilized language gets shored up, for example, in what is aptly termed hard-line feminist realism. Thinkers within this powerfully cast frame construe all of social life as a battleground, reversing the Clausewitzian dictum by viewing politics as war by other means. Women are advised to "fight dirty." Military metaphors are generously sprinkled about ("Who is the Enemy? Where is He Located?"). Politics is a case of oppressor/oppressed; and power is construed as crude instrumentality—a force to be deployed. When feminists reproduce this discourse, they shore up *power* understood in the singular—that is, as *le pouvoir*, meaning "force" or "superordination"—and thus can see politics *only* as a form of domination in which women might one day be equally "sovereign." Alternatively, to think of power in the plural, or as *des pouvoirs*, situates the analyst in such a way that she can recognize historic and current forms of female power(s)—and that is the route a more social, pacific feminism tends. Feminist realism, with its masculinist counterparts, magnifies and distorts one feature of a complex and by no means unambiguous truth, thereby shoring up rather than challenging the Western narrative of war and politics.

In its most tough-minded version, the woman leaps out of the fe-male/private side of the public/private divide basic to Machiavellian realism and straight into the arms of the hegemonic male: women can best him at his old game. In a variant on this theme, the male world is catalogued as a compendium of crimes against women, and men are apotheosized as the evil Other whose epitomizing act is to kill. With its symmetrical epitomizing act—to give birth—feminine nature comes to the rescue of the human race. Nancy Huston notes the "supreme irony, many feminist authors now describe childbirth in such a way as to assimilate it to war."[18] Examples include Marilyn French, in *The Women's Room*, writing that when you are pregnant, "You're like a soldier in a trench. . . . He gets to the point where he yearns for the battle [that is, the delivery of the child], even though he may be killed or maimed in it."[19] Phyllis Chesler, in *Women and Madness*, insists that "*all* women who bear children are committing, literally and symbol-ically, a blood sacrifice for the perpetuation of the species."[20]*

And in Monique Wittig's utopian feminist society of women fighters ("The women say, whether men live or die, they no longer have power"),[21] men are the enemy and midwives whoop when a child is born "in the manner of the women at war. This means that the mother has vanquished as a warrior, and that she has captured the child."[22] Quoting Mary Daly, one writer in this genre draws up her Manichean narrative: "The rulers of patriarchy . . . wage an unceasing war against life itself. . . . Women must understand that the female self is the enemy under fire from the patriarchy."[23] The rhetoric of war dominates; but, ultimately, the victims will vanquish their oppressors.

The triumph of the Beautiful Soul is anticipated in this alternative scenario. Rather than a stalemate in the sex war with men and women armed and *en garde*, fragments of Beautiful Soul imagery, in the strongest versions, foresee the triumph of the female principle, even-tually stopping all the evils traceable to masculinism (males are not Just Warriors but dangerous beasts in this narrative of binary opposites): environmental destruction, nuclear energy, wars, militarism, states.

The discourse of feminism's Beautiful Souls contrasts images of "caring" and "connected" females to "callous" and "disconnected"

---

* Note that this polemical assimilation of child*birth* to an imagined notion of the warrior *eager to kill* is very different indeed from the sacrificial structure of experience of soldiering and mothering I have discussed in chapter 6.

males. Men are the targeted enemy. One of the most extreme proposals flowing from this presumption is a call couched as point III of an agenda: "The proportion of men must be reduced to and maintained at approximately 10% of the human race." How we are to arrive at this perfect number is none too clear. But there is no doubt about who the agents of cultural warfare are: "It must be women in charge of the changes: woman-identified women, not women who are pawns of men, not women who out of their fear of losing their lives or those of their children, still hold to the securities of that dangerous patriarchal culture."[24] Matriarchal society, defined as a society "in which all relationships are modeled on the nurturant relationship between a mother and her child," is celebrated as the healing vision, the sealing over of all estrangements: womankind—coterminous with *human*—will be as one.[25] The trouble is: children are not grown-ups, and mothering is not and never has been a *wholly* beneficent activity.

Modified and more plausible—because less demonological and absolutist—versions of feminist realism and Beautiful Soul variations include, first, the current stance of the National Organization for Women. In 1981, NOW filed a legal brief as part of a challenge to all-male military registration. Beginning with the claim that compulsory, universal military service is central to the concept of citizenship in a democracy, NOW buttresses an ideal of armed civic virtue. If women are to gain "first-class citizenship," they, too, must have the right to fight. Laws excluding women from draft registration and combat duty perpetuate "archaic notions" of women's capabilities; moreover, "devastating long-term psychological and political repercussions" are visited upon women, given their exclusion from the military of their country.[26]

According to one critic, NOW's brand of equal-opportunity feminism here loses a critical edge, functioning instead to reinforce "the military as an institution and militarism as an ideology" by perpetuating the "notion that the military is so central to the entire social order that it is only when women gain access to its core that they can hope to fulfill their hopes and aspirations."[27] Because this contemporary discourse is entangled with presumptions geared historically against the values to which women are symbolically if problematically linked, feminist realism reiterates rather than deconstructs a model of armed civic virtue.

A second feminist vision, indebted to the discursive universe of

Beautiful Souls, shares just war's insistence that politics must come under moral scrutiny. Rejecting the hard-line gendered metaphysic of feminist Manicheism, the analyst in this mode nonetheless insists that women do have insights to bring to bear on the public world. Women are located as moral educators—*not* narrow and totalizing moralists— and as political actors. This discourse struggles to forge links between what Sara Ruddick has called "maternal thinking" and nonviolent theories of conflict without assuming that it is possible to translate easily maternal imperatives into a public good.[28] This view differs from celebrations of matriarchy that would structure all human relationships mimetically on the mother/child nexus. Its proponents recognize that people must grow up—to become citizens, among other things; and a civic being is not guaranteed nurture but is sure instead to find disagreement and conflict. In the knowledge that no unconditional mother love will come to the rescue, he or she must learn how to fight without destroying.

The feminist discourses I have just discussed and criticized frequently share attributes of that which they oppose: ahistorical abstractions; unreflective celebrations; taking a part of some complex situation, event, or idea and making it "the whole"; moralizing and dogmatizing. But it would be churlish to claim that these problematic features exhaust what there is to say about what women inspired either by "right to fight" or "no more war" positions—do *in the world*. I will look at the woman soldier in a moment, but here I want to note the transformative nature of the structure of experience women who have been "in residence" at Greenham Common (and its spinoffs) give witness to. Packing off for an indeterminate stay, some never before having left house, husband, and kids, many women found themselves for the first time taking risks and sharing discomforts with other women. Such action changed their lives, offering liberation from normal social constraints.

Costumes were in order at Women's Peace Camps, as were phantasmagorical makeup and hair styles. A romantic vision of the self as an identity in flux is central to all of this but hard to capture. Eric Leed notes: "The evanescent bonds and self-images formed in wars, revolutions, riots, carnivals, and New Year's parties are often historically invisible. They slip through the web of methods fashioned to describe the development of stable social and psychic entities."[29] Living on the boundary; existing on the edge. This is a seductive notion, and women who have been there—latter-day pilgrims and festival revelers acting out a morality play—say they will never be the same. This conviction must be treated with the same seriousness as the boundary experiences of war fighters; indeed, it has far more in common with those reported experiences than with the expectations that govern everyday life in Western societies.

Women as Warriors: "You're in the Army Now"

So what if I get my guts blown out all over the ship and it freaks the guys out? They would freak out if another guy had his guts blown all over too.

—LT. SHARON DISHER, project manager
Naval Civil Engineer Corps, United States Navy

Even as some women have turned Beautiful Soul assumptions into dramatic forms of anti-war protest, others have joined the camp of the warriors. The upshot is that the United States now has a higher percentage of women in its armed forces than does any other industrial nation, 10 percent of an overall force of 2.1 million.[30]* Although there is a long history of women in the services, the United States has abolished separate women's auxiliaries—WACs (Women's Army Corps), WAFs (Women in the Air Force), and so on—and set upon a course to "integrate" women into the regular army. The rise in the number of female soldiers is extraordinary. By the middle of 1948, the figures on women were down to approximately 8,000—about 1.4 percent of the total, given post-war demobilization. Now the armed forces are keen to deploy female "manpower" effectively, and run study after study to "address the effective utilization of female soldiers." Excluded

* Black women total nearly 30 percent of all enlisted women, a disproportionate number given the percentage of blacks in the population. Reasons given for black female enlistment include reshaping lives, trying to get away from "a dead-end existence," and the claim that the military "is the only institution in America that is truly mixed—racially integrated, male and female, grass roots and middle class."[31]

from combat, women will most likely not be able to "punch their tickets in the right places to be Chief of Staff," in the words of one observer. Nevertheless, those who favor the increase in female soldiers believe women are now being given a "chance to prove themselves," to "satisfy personal career goals and ambitions by moving up the ladder to senior management."[32]* States Carolyn Becraft of the Women's Equity Action League: "The United States is far and away ahead of other nations in making use of women in the military."[33]† There are those who oppose this move—among them General William Westmoreland, who believes women haven't got what it takes when it comes to combat; but it seems clear that women are now positioned in such a way that they will be *interior* to the action should a conventional war break out in Europe or in a number of other hot spots. And that's the way, on their own account, military women want it.

Although "optimum utilization" has yet to be achieved, this goal is nearer attainment each year. Put forward as a target of opportunity, the military has been transformed in modern American society into a career line. Older ideals of service, duty, and honor are shoved into the background. For example, an air force brochure designed to attract prospective men and women is entitled "The Changing Profession" and devotes its slick pages to extolling opportunities and listing possible career specialties. "Women compete equally," the air force says, noting that a "limited number of pilot, navigator and missile positions are made available on an annual basis."[34] The military has been "civilianized," brought under the same standards as consumerist civil society. It is another way to get ahead. The ideal of sacrifice has a musty, archaic air and extols an identity no longer honored even *within* the military itself (to overstate a strong current tendency).

Why has the United States moved with such rapidity to incorporate women? In West Germany, when the issue of enticing women to

---

* The so-called combat exclusion rule forbids sending women into combat; hence, they can neither fly warplanes nor serve aboard warships. Women are not issued "offensive" weapons, either. In a real war situation with the Russians, these distinctions would break down in practice, given the fuzziness of the distinction between offense and defense.

† Israel is often thought of as *the* developed country with women soldiers. But Israel exempts all married women from the military and from the reserves, and women of the Israeli Defense Forces have no combat duties on land or on sea. Opposition in Israel to putting women in the front lines is very strong.

maintain recruiting levels in the *Bundeswehr* came up, hackles got raised as men and women of the left and peace movements expressed their "outrage"—this despite an opinion survey in the spring of 1982 that showed 71 percent of women between the ages of eighteen and twenty-four to be in favor of women soldiers. The drive for sex equality propels some women into the ranks of the armed services in the United States, and the more determined and militant among them are frustrated at not being included in combat units. But that day can't be far away. Women in the armed forces are strongly in favor of it.[35] Despite the hopes expressed by observers like Betty Friedan—who assured readers of her *The Second Stage* (1981) that women warriors would, as women, have more sensitive concern for life than do male warriors, hence would be a force for caution and against brutality in any future war—such sentimentalism strains credulity.[36] Women soldiers do not speak that way. They are soldiers. Period.[37]

Many women, it seems, have chosen to identify with the soldiers' code, to mirror the male warrior—a vexing development if one's vantage point involves assaying the pitfalls as well as the possibilities of this individual and collective identity. Whatever life in the army may represent to the many women now involved, if one looks at the picture from a position outside such self-definitions one sees a male-forged identity being homogenized more widely. Yet, in light of the fact that all too many women are still prepared to have men *think* and *act* politically in their behalf, the woman soldier, if that is the way the woman has defined herself, seems not the worst possible alternative among those now available to us.

Here is the sort of evidence that invites this admittedly halfhearted conclusion: in a *Newsweek* poll conducted from 31 January to 1 February 1980 among 560 Americans aged eighteen to twenty-four, men and women opposed a return to the military draft by 51 percent to 44 percent overall. But when the responses were broken down by gender, more males opposed reinstitution of the draft (53 percent) than did females (47 percent), who were split down the middle on the question. But when the matter at hand was whether *young women* should be required to participate as well as young men should a draft become necessary, women respondents did an about-face with 39 percent agreeing, 58 percent saying no. The figures for men are a mirror image:

61 percent of young males saying yes, women should be drafted; 35 percent, no. An overwhelming percentage of men (77 percent) and women (73 percent) favor draft registration *for* men.[38]

Most salient are the figures on the draft, which suggest that many young women are prepared for the young men to fight rather than to participate themselves, given the former's approval of a male-only draft.* To me, this seems a piece of civic bad faith. Although I do not favor the draft, considering its history and the mobilization of the concept *service* (as being crafted to the purposes of nationalistic excesses by champions of civic unity), I have reluctantly come to the position that women should be granted no automatic exemption from this feature of civic life. Involuntary, unreflective pacifism is no pacifism at all.

I cling to no naïve liberal shibboleth that women if drafted in large numbers will transform the military and war fighting. I *know* the military will transform women. Whether one seeks to be self-identified in this way should be an issue with which young men and young women alike grapple: first, on the grounds of simple justice; and, second, as one way to relocate male and female selves to provide for a freer play of individual and civic capacities, in the hope of breaking the warrior/victim symbiosis.[39]†

Nor is this tentative conclusion the outgrowth of considerations derived from the abstract language of civic obligation; rather, it stems from a recognition that in the world we actually inhabit—rather than a dream world we would prefer—women will be drawn toward soldiering through conviction or circumstance. There seems little point in maintaining the pretense of combat exclusion for "their protection." Nobody can be protected any longer in the old sense of being "immune from possible destruction." But we have not yet found ways to talk about that recognition that might quicken rather than blunt civic democratic impulses.

* A *New York Times*–CBS News Poll, published 15 April 1986, also gives one pause. It shows that 68 percent of women and 56 percent of men are opposed to aid for the Contras: liberal opinion would take this as evidence of women's non-interventionist decency. But the figures seem blighted if one considers that *only* 25 percent of women knew that the United States supported the Contras as compared with double that figure for men. This suggests that women's civic opinions are less informed across the board about what United States policy actually is.

† On 25 March 1986, the first women crews of Minuteman 2 nuclear missiles took over command capsules at Whiteman Air Force Base, Montana. In a nuclear exchange, these women could be the first ordered to fire. Should we assume they will forbear because they are women?

Neither Warriors nor Victims: Men, Women, and Civic Life

Beyond War and Peace

RHETORICAL PRACTICES

> Nobody can be a citizen of the world as he is a citizen of his country. . . .
> No matter what form a world government with centralized power over
> the whole globe might assume, the very notion of one sovereign force
> ruling the whole earth, holding the monopoly of all means of violence,
> unchecked and uncontrolled by other sovereign powers, is not only a
> forbidding nightmare of tyranny, it would be the end of all political life
> as we know it. . . . A citizen is by definition a citizen among citizens of
> a country among countries.
> —HANNAH ARENDT, *Men in Dark Times*

Searching for a *language* in and through which to express the sentiments of civic life and the dangers and possibilities of the present
historic moment as these revolve around war and rumors of war, we
find many discourses clashing—strategic, psychological, apocalyptic—
and voices clamoring. The *strategic* voice is pre-eminently that of Dyson's warrior—cool, objective, scientific, and overwhelmingly male.
But more women aim to get in on this strategic enterprise, to certify
female voices as authoritative spokespersons for and of the world of
knowledge and power. The cool, in-command strategic voice talks in
the language of cost-benefit and command, and control and crisis
management. Women's Leadership Conferences on National Security
have been held since 1982 with the aim of devising "constructive ways
for women to become equal partners in the discussion and formulation
of national security policy." Panel discussions in recent years have
included such topics as "The Strategic Balance: Where Are We?"
Women, too, would speak in the voice of the knowing *insiders*. This
voice, as powerful as it is, has also been dissociated: that is, its linguistic
use of euphemisms (destroying villages as "pacification," for example,
or the MX missile as the "Peacekeeper") removes it from what it claims
to be talking about; moreover, strategic discourse, the preserve of
trained experts, is not available to most citizens, whether male or female, for the ordinary tasks of everyday civic life.

There is a less exclusive alternative language. Living, as we do, in
an era in which every issue quickly becomes one for therapy and gets

turned into a *psychological* problem, it is perhaps not surprising that nuclear weapons and war have been psychologized and that psychological discourse should proliferate on war and peace questions. Articles proliferate on "How to Deal with Despair," for psychological discourse claims that we must all feel dread in our current situation. If we are not suffering nuclear nightmares, this, too, is construed as proof that we are engaged in massive denial.

One way to treat such symptoms is to form a group, led by the man or woman best suited to facilitate group recognition of its own denials of the terror that all *must*, by definition, be feeling. This is called "An Integrated Large and Small Group Approach to Dealing with the Denial of Nuclear Danger." Nuclear support groups are urged on us all as a form of *emotional support—if* we do despair and empowerment exercises together—and of education. No doubt, some individuals are served through such devices. But when the psychological approach threatens to become just that—*the* approach—one must demur. It then invites individuals to become even more self-obsessed.

Contemporary nuclear-war discussions sometimes threaten to sink under the accumulated weight of pop psychologisms. The most dubious are analyses that deploy metaphors of illness, hence require calling upon experts who have the medicine and know-how to rid us of a deadly microbe or a "psychopathology." Only those who have expert knowledge of what ails us can be the doctors. The vast majority of us get construed as patients, not citizens. Drawing categories from individual psychology and putting them to work to cover complex structural realities and determinants sanctions an overpersonalizing of international political realities.

An example is Helen Caldicott, a medical doctor and leader in raising alarms about a nuclear disaster she finds imminent. She writes of "missile envy" as a psychopathology; of our whole planet as an organic entity that is "terminally ill, infected with lethal macrobes (nuclear weapons) that are metastasizing rapidly, the way a cancer spreads in the body." The problem with the cancer metaphor is that one cuts a cancer out or radiates it into oblivion. The equivalent presumably would be nuclear disarmament and destruction of stockpiled weapons. But when the analogy gets pressed between the structural realities of international politics and individual psychopathology, the inadequacies

246

of this discourse becomes clear. Women, once again, get dubbed with a rescue mission because they "innately understand conflict resolution," being "nurturers born with strong feelings for nurturing" given their anatomies and "hormonal constitutions"; males, having an excess of the "hormonal output of androgen," are more aggressive and bound to deploy deadly toys.[40]

This formulation does not seem terribly helpful. Males are always going to have an "excess" of androgen (it's an excess only if one sets up the female as a single human norm, which doesn't seem quite fair): that is part of what tilts a fetus toward maleness at about six weeks in the gestation period. So the upshot is that males need to be remade. No scheme that calls for the remaking of human nature has ever panned out. Indeed, it seems politically naïve—and the organic and psychological metaphors potentially dangerous in terms of the sorts of intervention they may invite.

The same holds for *apocalyptic* forebodings—assurances that we are doomed. Meant to warn us, books like Jonathan Schell's *The Fate of the Earth* paradoxically incite some among us to a fever pitch of expectation.[41] Among religious fundamentalists, for example, an anticipated nuclear holocaust is a source of joy, a sign that the Rapture is drawing nigh (that is, the divine rescue of true believers from the apocalyptic end): just before the earth gets devoured in an orgy of destruction, true believers will be lifted up and drawn to God as promised in the Book of Revelation. Apocalyptic warnings may be balm to the spirit of many, rather than a way to strike terror and in so doing to promote action. Central to the survivalist posture apocalyptic thinking promotes is a mode of reasoning also favored by contemporary hyperrealists—the "worst case" scenario; hence, the rhetorical ante gets upped.

Frustration and disillusionment, diffuse moral fervor rather than a civic commitment that requires patience stripped of utopian pretense, is the likely spinoff of apocalyptical discourse.

More promising is a *way of being* in the world that promotes civic identity and connection, even—at times, especially—if the form it takes is to reject the politics of the day or many of its central features. I have in mind the complex filiations of *private conscience* brought to bear on public and private lives and actions, offering tragic recognitions of necessary, even insoluble, conflicts, and, consequently, of limits to understanding and deed doing alike. Exemplary figures—an Addams or a Camus—help us get our own bearings in times of social conflict, not as characters to imitate but as persons whose struggles teach us much about our own. At odds with the all-or-nothing pronouncements of utopians and apocalyptic prophets, the articulators of limits create space for meaningful action by persons in the situations in which they find themselves.

Important, too, is the *ironic* spirit generated by the disillusioned of the First World War. The ironist locates himself or herself in the world in a way that mocks heroic words and action. The ironist knows that war is not what it is cracked up to be. Although the ironic voice is more likely to distance an individual from civic life than to locate him or her in the thick of things, unless we are attuned to that voice we will fail to understand why so many of the world's people are not as feverishly exercised over nuclear weapons as are we in contemporary American society. They know they have no control over events perpetrated by the superpowers. Ironic recognition of this historic limitation is the course most congenial to them.[42] Irony is less congenial to Americans because we do have far more power, hence more control, over the shape of things. We do have our own analogue of the ironists of the First World War in the words of many Vietnam vets who spend much of their time warning gung-ho young men that the actual experience of soldiering *must* fall short of the grand expectations many imagine.* We are torn between dangerous fantasies of perfect control—whether warlike or peaceful—and cynical withdrawal. Through ironic recognitions, however, we can deflect fanaticisms even as we sustain involvements.

We can also draw powerful insights from the discursive traditions assayed in earlier chapters. Realism, for example, is most compelling in insisting that "dirty hands" will always be a problem, and that diplomacy will always be necessary as a way to deflect the damage we might otherwise do to one another. The just-war–pacifist insistence that moral limits to political action must be articulated—not as an

---

* The irony of ironic discourse may be this: in order for such recognitions to bear the stamp of authenticity, perhaps the individual must participate in those experiences that most aptly lend themselves to this form of expression—soldiering being one such experience.

abstract set of formulae but as an ongoing feature of politics itself—can, and should, go hand in hand with realism's acknowledgment of the dilemma of political action. Just war and pacifist discourse also provide intimations that take us beyond, or through, *states*, helping us glimpse alternative ways in which collective political life may be expressed and structured. *If* that life is American, there are particular imperatives that unite and divide us, and that color the search for a civic language that takes us beyond the binary congealments of war and peace.

THE LESSON OF AMERICA: ABRAHAM LINCOLN AND MARTIN LUTHER KING, JR.

> No kind of greatness is more pleasing to the imagination of a democratic people than military greatness, a greatness of vivid and sudden luster, obtained without toil, by nothing but the risk of life.
> —ALEXIS DE TOCQUEVILLE, *Democracy in America*

No country can shed its historically determined skin. By this I mean that the strong pacifist appeal of a Gandhi would fall on uncomprehending ears should it be presented to an aggressively defined warrior band. There must be *within* a nation or a people a capacity, however unformed, that an appeal can forge into something palpable as an identity or discernible as an idea. Of American social identity it can be said that it is not only profoundly *individual*, it is also intensely linked to the belief that being an American means, or ought to mean, something. It means one has claims against any society that refuses to treat one with respect. It means America has stood historically for certain worthy ideals. A revitalized civic discourse must be neither abstract nor ahistorical but concrete and available to social participants. To break the war/peace nexus, such discourse must also incorporate potentially transformative features.

Liberal and republican elements commingle in this American civic stew, absorbing moments from Adam Smith's discussion of commercial society and the tempering of war in *An Inquiry into the Nature and Causes of the Wealth of Nations* (1776),[43] and imbibing some of Adam Ferguson's republican ferocity from *An Essay on the History of Civil Society*.[44] Published in 1767, Ferguson's work helped create what

249

J. G. A. Pocock calls "the Atlantic republican tradition." Ferguson bristles with the language of civic virtue *en garde*, noting that the "characters of the warlike and the commercial are variously combined," giving rise to war or inciting a desire of conquest, on the one hand, or leaving "people in quiet to improve their domestic resources," on the other.[45] He tracks a trajectory from primitive warriors to patriotic citizens; and, as J. G. A. Pocock points out, "Americans of that generation [the Revolutionary] saw themselves as freemen [to which I would add 'freewomen'], manifesting a patriotic virtue."[46] The second amendment to the U.S. Constitution speaks to the relation discerned between a popular militia and popular freedom in language that is geared toward civic freedom triumphant.

The discriminating values of the republican ideal (in which, remember, citizenship is, by definition, the property of a minority) evolved over time to a more solidly democratic ideal of citizenship. For many, the citizen-soldier was part of this ideal. For others, conscription constituted involuntary servitude as they brought liberal and libertarian beliefs into play to temper the republican ideal. Never easy with presumptions of the *primacy* of politics—and that is the direction civic republican virtue pushes—Americans modified its requirements additionally with Christian suspicion of most forms of political action, including the teaching that freedom *from* politics could also embody an ideal of a way of life.

From our site in the present, appeals to either old notions of a robust, coherent community *or* a robust, coherent individualist fall short of mapping the terms of a civic ethos that can continue to sustain us and form both personal and civic identities.* But there are, through the life and words of Abraham Lincoln, available to us recognitions that infuse a deep moral seriousness and an awareness of the inherent tragedy of political action.

Recently John Diggins has argued that Lincoln "helped heal that wound [the Machiavellian moment of severing politics from morality]

---

* *Pace* George Kateb's eloquent defense of *individualism* as the "most powerful idealism" and the one best suited to justify resistance in the face of possible nuclear extinction. Kateb believes this individualism can *fortify* the self in a task of "overcoming within themselves all that cooperates with the possibility of extinction." This extremely tall order is simply not within the reach of most persons in the heroic dimensions Kateb's argument requires.[47] Kateb forgets that resistance has been as much a corporate as an individual act.

by reintroducing into political discourse the Christian moralism that Machiavelli had purged from his theory of statecraft. Although there are elements of Machiavellian sensibility in Lincoln's conviction that he, too, was experiencing a universe eclipsed in moral darkness . . . , Lincoln was convinced that ultimate moral questions did not admit of relativistic interpretations," even though they might require trimming one's actions and going for less than one's pure convictions would insist upon.[48] Slavery *is* wrong. But ridding a nation of it involves a series of tragic encounters and compromises.

Lincoln instructs us today, chastening the arrogance of rationalist realists and strategic-deterrence model builders in his confession: "I have not controlled events." Rejecting Machiavellian instrumentalism and the "logic of success," Lincoln hoped more for the goodness of the [nation's] soul than for the greatness of the state. Thus, grand visions of wholesale *civic unity* fall through the cracks of Lincoln's recognitions. A nation must learn how to endure defeats—better that than how to crow in victory and war-bought unity. A people is well advised to "renounce heroically the very things that enslave the spirit," rather than to seize aggressively at a greatness that corrodes its soul. The Civil War was not fought between one side that was righteous and another that was not; rather, "Both read the same Bible, and pray to the same God. . . . But let us not judge, that we be not judged."[49] Diggins may be right—this is a remarkable stance for a political leader; but it is *the* stance of our greatest political leader, one over and over reinscribed on the American conscience, hence continuing to shape our political consciousness.* It means that *power* and *interests* alone cannot and need not define our politics. We can, with Lincoln, open our hearts to pity, recognizing that although there is no redemption through war, there is the possibility of redeeming a defeat—and a victory—but only if we chasten the nihilistic disillusionment or blind triumphalism each invites; only if we put away the grand narrative of armed civic virtue.

As I write these words, we Americans are in the midst of a period of popular enthusiasm that many choose to call the "new patriotism."

---

* Adam Michnik's essay "Maggots and Angels" is a powerful piece of anti-Manichean writing about the complexities of Polish history and politics that refuses to locate pure virtue and pure evil in individuals, movements, or sides.[50]

Much of it looks suspiciously like the old nationalism to me, an aggressive self-identity that courts arrogance through our identification with the state's awesome preserve of force and invites dreams of a perfectly unified society, highly mobilized and ready to do battle. The language of nationalism is like the language of war, drastically oversimplified, deployed, in Orwell's words, either in "boosting or in denigrating," classifying whole peoples as "good" or "bad."[51]*

Patriotism, however, is part of our repertoire of civic ideals and identities. While its excesses may be lamented, it cannot, and should not, be excised, for patriotism also taps love of country that yields civic concern *for* country. Attached more to the sense of a political and moral community than to a state, patriotism can be, and has been, evoked to bring out the best in us—even as, when it shades into nationalism, it can arouse the worst.

For example, we have a sense of what is at stake to describe Martin Luther King, Jr., as a "great American patriot" who had the courage to criticize his "beloved country" in time of war even as he affirmed his ties to the country. It would make no sense at all to think of King as a "great American nationalist." His "I have a dream" speech (delivered at the 1963 March on Washington) is of a vision of a multinational, racially and regionally diverse, plural democracy. In his "Beyond Vietnam" speech of 4 April 1967, King makes a plea "to my beloved nation. This speech is not addressed to Hanoi or to the National Liberation Front. It is not addressed to China or to Russia . . . but rather to my fellow Americans who, with me, bear the greatest responsibility in ending a conflict that has exacted a heavy price on both continents." These are the words of a patriotic citizen: for nationalist concepts of citizenship turn upon the construction of foreign enemies as implacable, demonized foes.

We can be patriots within the diverse frames this book has unraveled and restitched into a civic fabric that might deflect violence without presuming its ultimate defeat. But it is a *chastened* patriot I have in mind, men and women who have learned from the past. Rejecting counsels of cynicism, they modulate the rhetoric of high patriotic purpose by keeping alive the distancing voice of ironic remembrance and

* For example, nearly half the American people, according to a *New York Times* poll conducted in 1985, believe that Americans care more about their children than Russians do about theirs.[52]

252

recognition of the way patriotism can shade into the excesses of nationalism; recognition of the fact that patriotism in the form of armed
civic virtue is a dangerous chimera. The chastened patriot is committed
and detached: enough apart so that she and he can be reflective about
patriotic ties and loyalties, cherishing many loyalties rather than
valorizing one alone.

A civic life animated by chastened patriotism bears implications for
how we think of peace and war, and for the pitfalls in how each has
been construed.

THE PROBLEM WITH PEACE

> But He, her fears to cease,
> Sent down the meek-eyed Peace;
> She, crown'd with olive green, came softly sliding
> Down through the turning sphear,
> His ready Harbinger,
> With Turtle wing the amorous clouds dividing,
> And waving wide her mirtle wand,
> She strikes a universall Peace through Sea and Land.
>   —MILTON, "Ode on the Morning of Christ's Nativity"

Peace is an ontologically suspicious concept, as troubling in its own
way as war. War's historic opponents—those who want peace—are
inside a frame with war. These two, peace and war, help structure
Western civil society's view of itself, with protests against war couched
in terms that mirror that which they oppose: peace to war; anti-bellicist
femininity to bellicose masculinity; harmony to disorder; homogeneity
to heteronymy; and so on. Peace cannot exist without war, and both
are problematic notions, obsolete in an era of nuclear weaponry, constantly reduced launch-on-warning time, spy satellites, and national
defense as a potential form of civic suicide.

War as it has been historically constituted is an artifact of the grand
narrative I have traced—just as peace, emerging as the antinomy of
war, got similarly constructed. This notion may help explain why so
many Americans favor peace but do not support peace movements.
They do not understand what this condition called "peace" is supposed
to be: The absence of armed conflict? Perpetual world harmony in
which communist lions lie down with capitalist lambs (or the other

way round depending upon where one situates one's animal meta-phors)? A world order of some sort? An end to the need for any defense? Peace seems bland, an unattainable absolute, a vision of perpetual har-mony that jars with common sense. Given the linkup of peace to "security" and "order," these reservations must be taken seriously. In the eighteenth century, peace was in part a dream of national discipline and a pacified population.

Too often perfect harmony or ongoing blissfulness capture dreams of peace. In the 1985 film *Plenty*, for example, the female protagonist's life goes downhill once peace comes, once the excitement of the Second World War is over. She cannot recapture the thrill. We understand why fully when the film ends on a flashback. Peace has come. She, a young Englishwoman working with the French Resistance, runs to the top of a hill. It is a beautiful sunny day. The French countryside stretches before her, glorious, shimmering in the sun; and she cries, now that peace has come, there will be "days like this" without end. Her utopian ontology of peace, her yearning for continuing epiphany, guarantees her subsequent cynicism and hopelessness.

Sometimes dreams of peace are more modest. Robert Graves notes that he and Siegfried Sassoon, during the First World War, defined war "in our poems by making contrasted definitions of peace. With Siegfried it was hunting and music and pastoral scenes; with me it was chiefly children."[53] Graves's homey antithesis to war was, to him, within reach. But most of the time "peace" promises more than it can deliver. One of Studs Terkel's respondents said that the Second World War led people to "expect more. People's expectations . . . were raised. There was such a beautiful dream. We were gonna reach the end of the rainbow. When the war ended, the rainbow vanished. . . . There was no peace, which we were promised."[54]

Starkly, J. Glenn Gray notes that peace may expose in people a "void" that "war's excitement" had enabled them to keep covered up. People fear a "sterile peace," a world without liminal enthusiasms and possibilities.[55] They endorse war as the opposite of peace, according to Eric Leed, in part because they yearn for a mode of existence that differs "markedly from bourgeois society." The First World War promised such release. Paradoxically, out of treatment of the "shell-shocked" in that war emerged a model of the self that predominates for psychological theorizing in modern Western society: the ideal of a unified self with a solid identity. Erik Erikson constructed his notion of identity specifically as a "corrective to war" in which people experience themselves as not "the same," as changed.[56]

But perhaps a presumption of self-sameness is itself one of the burdens from which wartime has traditionally posed relief. We usually think of war as a threatening disorder; for many people, war is a promise, a way to break out of the social atomization, contractual relationships, and cramped analytic attitudes characteristic of modern society.* War has promised this, and sometimes delivered it, to men *and* women. But the grand project of peace *is* false in presuming the possibility of attaining and maintaining a project of equilibrium, of order, of harmony.

The title of Immanuel Kant's great essay tells the story: "Perpetual Peace." The key, for Kant, is *perpetual*; indeed, to modify peace by "perpetual" constitutes a pleonasm. Peace that is not perpetual is a mere truce. A genuine peace treaty must nullify *all* existing causes of war. There must be *no mental reservation whatever*. War is absolutely prohibited. Peace derives from republican civil constitutions; its origin is pure—the pure concept of right. A league of such purely constituted republics will end war forever. Kant's grand design is laced through and through with a grand teleology, willed by Nature for She has set up all the preconditions, using even war to gain the absolute of perpetual peace. Although current analysts sometimes think of a "Kantian type of future," that future must forever elude us—but Kant's notion of a moral absolute will no doubt forever haunt us. It is a ghost that should be put to rest.

His peace is a solipsistic dream which can "exist among like kinds and equals" only, making of the mere existence of "otherness" a flaw in the perfect scheme of things. Kantian peace promises not only what can never be but what would be undesirable in any case, a logic that cannot get beyond the logic of war, conjuring up images of "two irreducible opposites confronting one another," with war the enemy to peace. And politics, which is the way human beings have devised for dealing with their differences, gets eliminated.[58]

---

* Hannah Hafkesbrink, a German feminist, says that many of her generation felt the war had released them from "an economic, materialistic, and technological order of existence," making the war a "moral project, in contrast to the amorality of the marketplace."[57]

BREAKING THE DEADLOCK (OF WAR'S MOBILIZED LANGUAGE)

> But what does it mean to think in wartime images? It means seeing ev-
> erything as existing in a state of extreme tension, as breathing cruelty and
> dread. For wartime reality is a world of extreme, Manichaean reduction,
> which erases all intermediate hues, gentle, warm, and limits everything
> to a sharp, aggressive counterpoint, to black and white, to the primordial
> struggle of two forces—good and evil. Nothing else on the battlefield!
> Only the good, in other words, us, and the bad, meaning everything that
> stands in our way, which appears to us, and which we lump into the
> sinister category of evil.
>   —RYSZARD KAPUSCINSKI, "1945"

Wartime's mobilized language is infused within the metaphors and tropes of everyday discourse. We are weaned on such opposites as good versus evil, peace versus war, just versus unjust. To deflect this warlike way of thinking is impossible so long as we remain enthralled by grand teleologies of historic winners and losers; so long as our reigning narratives are triumphalist accounts of our victories; so long as our identities are laced through with absolutist moralisms; so long as we seek, need, or require, *on this earth*, a unifying experience that total war or perpetual peace alone promises. To appreciate the relativity of all antagonisms and friendships, to see in others neither angels nor demons, puts one on a track different from that laid down by those who would organize and systematize reality in a relentlessly total way. In wartime we see this logic at its most unlovely, as slackers, enemies within, all undigested bits are compelled to conform or are severed from the body politic.

The nation, one and indivisible, requires an external enemy to come into being. Such is Hannah Arendt's claim, and she finds a dream of unity endemic to the ideal of all modern nationalisms. Generative vi-olence in which individuals and whole nations find themselves in a mimetic relationship—through which one keeps alive the danger by which one feels threatened, through which one requires an enemy— frequently takes hold of the political imagination. Historically war has crystallized and unified a nation's sense of self for, in and through war, diverse peoples have entered into relationships by pitting them-selves *against* one another and, in so doing, distinguishing themselves

256

for what they are by distancing themselves from what they are not.

Once a nation has defined itself through war, the habit may get hard to break.* Encoded within the politics of the state, such war definitions reinforce the pretensions of sovereignty and reinscribe calls to national unity. It is difficult to stem the tide. But that, too, may lie within our reach if we recognize that the state is not coterminous with the political. Sheldon Wolin argues that the state's historic *raison d'être* "was to develop, or better, to capitalize the power of society—the power resident in the human activities, relationship, and transactions that sustain life and its changing needs. The state became a coercive agency, declaring and enforcing law, punishing miscreants ..., systematizing taxation, ... waging war, and seeking empire."[59] This agency works to foreclose alternatives.

But politics is still possible, even a politics of grand enthusiasms that Paul Virilio calls "holy *non* war": politics that confronts extremes but avoids sacralizing violence.[60] This politics beyond war and peace refuses to see *all* right and good on one side only. Its ethos offers values for which one might die but not justifications of the need to kill, on the assumption that one cannot arrogate this absolute unto oneself. The adherents of this politics are those who share a moral humility that repudiates the fiction that any means are possible if one declares one's ends to be good or just.[61] To create social space through experiments in action with others would serve to free up identities, offering men and women the opportunity to share risks as citizens, to take up nonviolence as a choice, not a given. Collective violence would remain—as the way men and women who see themselves as guardians of the state undertake their task and forge their identities; but it would no longer play the same role as its grand narrative got repeatedly challenged, not by "peace" but by politics.

Open to foreignness and differences from a mood that embraces less purity, hence sees less danger in others; that can practice "live and let live"—that is the intimation to which these reflections on women and war lead. Is it possible for human beings to accept life as a risk-taking adventure filled with uncertainty? Certitudes invite the arrogance of nationalistic excess and structure solemn, brittle identities—the tough

* Look, for example, at Britain's "born again" nationalism during the Falklands/Malvinas war.

male warrior; the pure, pacific woman—that not only keep others out, and construe them as enemies, but that preclude any inner dialogue with one's own "others."

For the political embodiment of the attitude I here suggest, I return to the chastened patriot. He or she has no illusions: recognizing the limiting conditions internal to international politics, this civic being does not embrace utopian fantasies of world government or total disarmament. For neither the arms-control option (as currently defined) nor calls for immediate disarmament are bold: the first, because it is a way of doing business as usual; the second, because it covertly sustains business as usual by proclaiming solutions that lie outside the reach of possibility.

Devirilizing discourse, in favor not of feminization (for the feminized and masculinized emerged in tandem and both embody dangerous distortions) but of politicization, the chastened patriot constitutes men and women as citizens who share what Hannah Arendt calls "the faculty of action." This citizen is skeptical about the forms and claims of the sovereign state; recognizes the (phony) parity in the notion of *equally* "sovereign states," and is thereby alert to the many forms hegemony can take; and deflates fantasies of control. Taking a cue from Arendt, this citizen gives "forgiveness" a central role as one way human beings have to break cycles of vengeance. *Ascesis*—a refraining or withholding, a refusal to bring all one's force to bear—surfaces, in this vision of things, as a strength not a weakness.

As Americans, citizens of as a strong and dominant nation of awesome potential force, we are invited to take unilateral initiatives in order to break symbolically the cycle of vengeance and fear signified by that very force. As individual men and women we are invited to examine, and take up, the alternatives, woven throughout the story of women and war, to identities that lock us inside the traditional, and dangerous, narrative of war and peace. The dream I am dreaming as I end these reflections is not one of solemn deed doers but of zestful act takers, experimenting with new possibilities playfully but from a deep seriousness of purpose.

NOTES

Introduction: Beautiful Souls/Just Warriors:
The Seduction of War

1. See the collected essays of Clifford Geertz, *The Interpretation of Culture* (New York: Basic Books, 1973).

2. G. W. F. Hegel, *The Phenomenology of Spirit*, trans. A. V. Miller (Oxford: Clarendon Press, 1977), pp. 399–400.

3. See Emma Goldman, *The Traffic in Women and Other Essays in Feminism* (New York: Times Change Press, 1970), passim.

4. Quoted in Elizabeth Cady Stanton, Susan B. Anthony, and Matilda Joslyn Gage, eds., *History of Woman Suffrage*, vol. 1 (Rochester, N.Y.: Charles Mann, 1889), p. 145.

5. Susan B. Anthony and Ida Husted Harpers, eds., *History of Woman Suffrage*, vol. 5 (New York: J. J. Little & Ives, 1922), p. 585.

6. Stanton, Anthony, and Gage, *History of Woman Suffrage*, vol. 2, pp. 351–52.

7. Kurt Vonnegut, *Slaughterhouse Five* (New York: Delacorte Press, 1969).

8. Leila J. Rupp, *Mobilizing Women for War* (Princeton, N.J.: Princeton University Press, 1978), p. 180.

9. D'Ann Campbell, *Women at War with America* (Cambridge, Mass.: Harvard University Press, 1985).

10. Peter Berger, "On the Obsolescence of the Concept of Honor," in Michael J. Sandel, ed., *Liberalism and Its Critics* (Oxford: Basil Blackwell, 1984), pp. 149–58.

11. John Keegan, *The Face of Battle* (New York: Penguin Books, 1983), pp. 328, 331.

12. John Koten, "Too Tough a Plane? F-16 Can Take Stress That Its Pilot Can't," *Wall Street Journal*, 28 May 1986, p. 1.

13. J. Glenn Gray, *The Warriors* (New York: Harper Colophon, 1970), p. 217.

14. Georgia Dullea, "Women Who Served in Vietnam Emerge as Victims of War Strain," *New York Times* (23 March 1981, pp. 1, B-12).

15. Nancy Huston, "Tales of War and Tears of Women," *Women's Studies International Forum* 5 (3/4 [1982]):271–82; quotation on p. 280.

16. As told by Wendy Chapkis and Mary Wings, "The Private Benjamin Syndrome," in *Loaded Questions: Women in the Military*, ed. Wendy Chapkis (Amsterdam: Transnational Institute, 1981), pp. 17–21.

17. Ibid., p. 18.

18. Adam Clymer, "What Americans Think Now," *New York Times Magazine* (31 March 1985, p. 34).

Notes

Chapter 1. Not-a-Soldier's Story:
An Exemplary Tale

1. Ernie Pyle, *Here Is Your War* (New York: Henry Holt, 1943), p. 102.

2. Ibid., pp. 212–13.

3. Norman Mailer, *Why Are We in Vietnam?* (New York: Holt, Rinehart & Winston, 1967), pp. 85, 208.

4. From *Echoes of the Great War: The Diary of the Reverend Andrew Clark 1914–1919*, ed. James Munson (New York: Oxford University Press, 1985); the epigraph quote is on p. 11.

5. Poetry extracts by Francis Finch and Virginia Frazer Boyle were copied verbatim into my journal.

6. Mohandes K. Gandhi, *Autobiography or The Story of My Experiments with Truth* (Ahmedabad: Navajivan, 1940).

7. James Reston, Jr., *Sherman's March and Vietnam* (New York: Macmillan, 1984), p. 173.

8. William L. Shirer, *The Rise and Fall of the Third Reich: A History of Nazi Germany* (New York: Simon & Schuster, 1960).

9. See, for example, Sarah Gordon, *Hitler, Germans and the "Jewish Question"* (Princeton, N.J.: Princeton University Press, 1984).

10. Michael Sandel, *Liberalism and the Limits of Justice* (Cambridge: Cambridge University Press, 1982), p. 183.

11. *New York Times*, 7 May 1981, p. C2.

12. George Eliot, cited in Ruby V. Redinger, *George Eliot: The Emergent Self* (New York: Alfred A. Knopf, 1975), p. 315.

Chapter 2. The Discourse of War and Politics:
From the Greeks to Today

1. Arlene Saxonhouse, "Men, Women, War and Politics: Family and Polis in Aristophanes and Euripedes," *Political Theory* 8 (1 [February 1980]):65–81; quotation on p. 66.

2. My analysis differs in several important ways from that proffered by Nancy Hartsock in *Money, Sex and Power* (New York: Longmans, 1984).

3. Oswyn Murray, "Literature and War: The Greeks," *Times Literary Supplement*, 17 May 1985, p. 546.

4. See Nancy Huston, "Tales of War and Tears of Women," *Women's Studies International Forum* 5 (3/4 [1982]):271–82.

5. Dennis Porter, "Response to Jean Bethke Elshtain's 'From Machiavelli to Mutual Assured Destruction: Reflections on War and Political Discourse,' " *Occasional Papers*

Notes

II (Amherst, Mass.: Institute for Advanced Study in the Humanities, 1985), pp. 35–42.

6. Simone Weil, *The Iliad, or the Poem of Force* (Wallingford, Pa.: Pendle Hill, 1956).

7. Huston, "Tales of War," p. 276.

8. Hannah Arendt, *The Human Condition* (Chicago: University of Chicago Press, 1958), p. 19.

9. Steven Salkever, "Women, Soldiers, Citizens: Plato and Aristotle on the Politics of Virility" (paper given at 1983 meeting of the American Political Science Association).

10. Arlene W. Saxonhouse, "Aeschylus' *Oresteia*: Misogyny, Philogyny and Justice," *Women and Politics* 4 (2 [Summer 1984]):11–22.

11. The interested reader may want to consult Sarah B. Pomeroy's *Goddesses, Whores, Wives and Slaves: Women in Classical Antiquity* (New York: Schocken Books, 1976); and Alasdair MacIntyre, *A Short History of Ethics* (New York: Macmillan, 1966) for additional background, historical and philosophical, on this period and the transition to settled polis life.

12. See George Steiner, *Antigones* (New York: Oxford University Press, 1984), p. 183; and my much more modest effort, "Antigone's Daughters," *Democracy* (April 1982):46–59, specifically p. 53.

13. The story of a later female warrior is told by Marina Warner in *Joan of Arc: The Image of Female Heroism* (New York: Vintage Books, 1982).

14. Aeschylus, *The Oresteia*, trans. Robert Fagles (New York: Penguin, 1977), p. 271.

15. Ibid., p. 269.

16. Hartsock, *Money, Sex and Power*, p. 191.

17. Robert Fagles, introduction to Aeschylus, *Oresteia*, p. 74.

18. Hartsock, *Money, Sex and Power*, p. 206n19.

19. Aeschylus, *Oresteia*, p. 264.

20. David Grene and Richard Lattimore, eds., *Greek Tragedies*, vol. 2 (Chicago: University of Chicago Press, 1960), pp. 272, 288 (from *The Trojan Women*).

21. Saxonhouse, "Men, Women, War and Politics," pp. 74–75.

22. Plato, *The Republic* 6.503D, trans. Allan Bloom (New York: Basic Books, 1968).

23. Plato, *Republic* 3.416E.

24. Ibid., 6.503D.

25. Aristotle, *Politics*, ed. and trans. Ernest Barker (New York: Oxford University Press, 1962), Bk. 2, p. 79.

26. Ibid., Bk. 6, p. 245.

27. Ibid., Bk. 7, pp. 285, 317.

28. Ibid., Bk. 1, p. 5.

29. Jacob Burckhardt, *The Civilization of the Renaissance in Italy*, vol. 1 (New York: Harper Torchbook, 1985), cites Machiavelli in discussing "war as a work of art." (Originally published in 1860.)

30. Alexis de Tocqueville, *Democracy in America*, vol. 2 (New York: Vintage Books, 1945), p. 247.

31. Niccolò Machiavelli, *The Discourses* (New York: Modern Library, Random House, 1950), p. 104.

32. Jeff Weintraub, "Virtue, Community, and the Sociology of Liberty: The Notion

Notes

of Republican Virtue and Its Impact on Modern Western Social Thought" (Ph.D. diss., University of California, Berkeley, 1979).

33. Machiavelli's *Discourses*, cited in J. R. Hale, "Machiavelli and the Self-Sufficient State," in *Political Ideas,* ed. David Thomson (New York: Penguin Books, 1966), pp. 22–33; quotation on p. 32 (from Christian E. Detmold translation of *The Discourses*).

34. J. G. A. Pocock, *The Machiavellian Moment* (Princeton, N.J.: Princeton University Press 1975), p. 210.

35. Ibid., p. 213.

36. See my discussion in *Public Man, Private Woman: Women in Social and Political Thought* (Princeton, N.J.: Princeton University Press, 1981), pp. 92–99; and Hanna Fenichel Pitkin, *Fortune Is a Woman: Gender and Politics in the Thought of Niccolo Machiavelli* (Berkeley: University of California Press, 1984), for an extended treatment of Fortuna.

37. Machiavelli, *Discourses*, p. 489.

38. Augustine, *The City of God*, ed. David Knowles (Baltimore, Md.: Penguin Books, 1972).

39. Jean-Jacques Rousseau, *On the Social Contract with Geneva Manuscript and Political Economy*, ed. Roger D. Masters (New York: St. Martin's Press, 1978), p. 128.

40. Ibid., p. 163.

41. Ibid., p. 202.

42. Jean-Jacques Rousseau, *The First and Second Discourses* (New York: St. Martin's Press, 1964), p. 55 (from the *First Discourse*).

43. Jean-Jacques Rousseau, *The Government of Poland* (Indianapolis: Bobbs-Merrill, 1972), p. 17.

44. Plutarch, *Moralia* III, 459, 463, trans. Frank Cole Babbitt (Cambridge, Mass.: Harvard University Press, 1931).

45. Hannah Arendt, *On Revolution* (New York: Penguin Books, 1977), pp. 97, 99.

46. Robespierre's speech is reproduced in Michael Walzer, *Regicide and Revolution* (Cambridge: Cambridge University Press), pp. 178–94.

47. See Michel Foucault, *Discipline and Punish* (New York: Vintage Books 1979), pp. 168–69.

48. Carnot, a former army officer, became a member of the Committee of Public Safety in 1793. His background suited him for the task of mobilization.

49. Part of the text of the 1789 Declaration of the Rights of Man appears in Gerald P. Dartford, *The French Revolution* (Wellesley Hills, Mass.: Independent School Press, 1972), p. 92.

50. Michael Howard, *The Causes of War* (Cambridge, Mass.: Harvard University Press, 1984), pp. 159–60.

51. See Eugene Weber, *Peasants into Frenchmen: The Modernization of Rural France: 1870-1914* (Stanford: Stanford University Press, 1976), p. 293.

52. Ibid., p. 296.

53. Pierre-Jakez Hélias, *The House of Pride: Life in a Breton Village*, trans. June Guicharnaud (New Haven: Yale University Press, 1978), pp. 30–31.

54. Ibid.

55. Jean-Jacques Rousseau, *Politics and the Arts: Letter to M. D'Alembert on the Theatre*, trans. Allan Bloom (Ithaca, N.Y.: Cornell University Press, 1960), p. 106.

56. Rousseau, *Government of Poland*, p. 19.

Notes

57. Jean-Jacques Rousseau, *Emile*, trans. Allan Bloom (New York: Basic Books, 1979), pp. 37–38*n*.

58. Rousseau, *Government of Poland*, p. 87.

59. Steiner, *Antigones*, p. 149.

60. Mary Wollstonecraft, *A Vindication of the Rights of Woman* (New York: W. W. Norton, 1967), p. 75. Portions of this discussion draw on Jean Bethke Elshtain, *Meditations on Modern Political Thought* (New York: Praeger, 1986), pp. 37–54.

61. Rousseau, *Emile*, p. 40.

62. Plutarch, *Moralia* III, pp. 461–63.

63. Wollstonecraft, *Vindication*, p. 43.

64. Ibid., p. 287.

65. See "High Court Backs Military Ban on the Yarmulke," *New York Times*, 30 March 1986, p. 8E.

66. See, for example, Arendt, *On Revolution*, pp. 108–9.

67. The most important Hegel text on these and matters to follow is the *Philosophy of Right* (1821), trans. T. M. Knox (London: Oxford University Press, 1973). Hegel's *Phenomenology of Spirit* (1807, trans. A. V. Miller [Oxford: Oxford University Press, 1977]) is vital to an understanding of the terms of forging identity through conflict and combat with and against the Other. Shlomo Avineri, *Hegel's Theory of the Modern State* (Cambridge: Cambridge University Press, 1974), and Charles Taylor, *Hegel* (Cambridge: Cambridge University Press, 1975), are indispensable commentaries.

68. See also Hegel's discussion of Antigone in the *Phenomenology* and *Philosophy of Right* and contrast it with Steiner's *Antigones*.

69. Hegel, *Philosophy of Right*, p. 210.

70. Ibid., pp. 211, 212.

71. Ibid., p. 278.

72. Howard, *Causes of War*, pp. 103–4.

73. Bernard Brodie, "A Guide to the Reading of *On War*," in Carl von Clausewitz, *On War*, ed. and trans. Michael Howard and Peter Paret, with essays by Paret, Howard, and Brodie (Princeton, N.J.: Princeton University Press, 1984), pp. 641–711.

74. Clausewitz, *On War*, pp. 137–38.

75. Ibid., p. 138.

76. See the somewhat dated but still valuable book by Raymond L. Garthoff, *Soviet Military Doctrine* (Glencoe, Ill.: Free Press, 1953), especially pp. 52–56 on Clausewitzian rumblings.

77. Marie von Clausewitz, preface to *On War*, pp. 65–67.

78. Ibid.

79. Ibid.

80. Hannah Arendt, *On Violence* (New York: Harcourt, Brace, Jovanovich, 1969), pp. 35, 56.

81. Quotations are drawn from Diane Paul's essay "In the Interests of Civilization: Marxist Views of Race and Culture in the Nineteenth Century," *Journal of the History of Ideas* 42 (January/March 1981):115–38.

82. See discussion in V. I. Lenin, *State and Revolution* (New York: International Publishers, 1974).

83. Sigmund Freud, "A Weltanschauung?" Lecture XXXV of *New Introductory*

Notes

Lectures on Psychoanalysis, The Standard Edition of the Complete Psychological Works of Sigmund Freud, vol. 22 (London: Hogarth Press, 1964), p. 181.

84. Lenin, *State and Revolution,* pp. 47, 55, 20.

85. "Terrorism and Communism," *Spartacist* 17–18 (August–September 1970):4–7; quotation on p. 7.

86. See the discussion in Elshtain, *Public Man, Private Woman: Women in Social and Political Thought,* pp. 256–67.

87. Mao Tse-tung, *On Protracted War* (Peking: Foreign Languages Press, 1966), p. 57.

88. Mao Tse-tung, *Problem of War and Strategy* (Peking: Foreign Languages Press, 1965), pp. 8–9.

89. Arendt, *On Violence,* p. 11.

90. Jean-Paul Sartre, introduction to Frantz Fanon, *The Wretched of the Earth* (New York: Grove Press, 1981).

91. Cited in Mary Midgley, *Wickedness: A Philosophical Essay* (London: Routledge & Kegan Paul, 1984), p. 195. Cf. Ronald N. Stromberg, *Redemption by War: The Intellectuals and 1914* (Lawrence, Kan.: Regents Press, 1982).

92. The central text is Thomas Hobbes, *Leviathan* (New York: Collier Books, 1966), especially chap. 13.

93. Hobbes, cited in Arendt, *On Violence,* p. 5.

94. Donald W. Hanson, "Thomas Hobbes's 'Highway to Peace,'" *International Organization* 39 (2 [Spring 1984]):329–54.

95. See Stanley Hoffman, "International Relations. The Long Road to Theory," *World Politics* 11 (3 [April 1959]):346–77.

96. An incisive analysis is offered by Bradley S. Klein in "Strategic Discourse and its Alternatives" (unpublished manuscript, 1986). The words in quotes are drawn from Klein's paper.

97. Arendt, *On Violence,* p. 12.

98. See James Der Derian, *On Diplomacy: A Genealogy of Estrangement* (Oxford: Basil Blackwell, 1987).

99. Howard, *Causes of War,* pp. 139–40.

Chapter 3. Exemplary Tales of Civic Virtue

1. Michel Foucault, ed., *I, Pierre Rivière, having slaughtered my mother, my sister and my brother...* (Lincoln: University of Nebraska Press, 1983), p. 200.

2. C. Vann Woodward and Elisabeth Muhlenfeld, eds., *The Private Mary Chesnut: The Unpublished Civil War Diaries* (New York: Oxford University Press, 1984), p. 21.

3. Ibid., pp. 202, 243.

4. Ibid., pp. 78–79.

5. Ibid., p. 252.

6. Matthew Page Andrews, *The Women of the South in War Times* (Baltimore: Norman, Remington, 1920), pp. 21, 76, 79.

Notes

7. Quoted in J. M. Taylor, *Eva Perón: The Myths of a Woman* (Chicago: University of Chicago Press, 1979), p. 128.

8. Andrews, *The Women of the South*, pp. 112–13.

9. Francis Butler Simkins and James Welch Patton, *The Women of the Confederacy* (Richmond and New York: Garrett & Massie, 1936), p. 17.

10. Ibid., pp. 18, 14.

11. Ibid., p. 81.

12. James Reston, Jr., *Sherman's March and Vietnam* (New York: Macmillan, 1984), pp. 92–93.

13. Ibid., p. 117.

14. Frank Moore, *Women of the War: Their Heroism and Self-Sacrifice* (Hartford, Conn.: S. S. Scranton, 1867), pp. iv–v.

15. Ibid., p. 17.

16. Rita Mae Brown, *High Hearts* (New York: Bantam Books, 1986).

17. Moore, *Women of the War*, pp. 17, 36.

18. Ibid., pp. 55, 56, 86.

19. Ibid., p. 529.

20. Herbert Fairlie Wood and John Sweetenham, *Silent Witnesses* (Toronto: Hakkert, 1974), p. 5.

21. Cited in D. E. D. Beales, "Mazzini and Revolutionary Nationalism," in David Thomson, *Political Ideas* (New York: Penguin Books, 1982), pp. 143–53; quotation on p. 146 (from Mazzini's *The Duties of Man*).

22. Emile Durkheim, *Moral Education. A Study in the Theory and Application of the Sociology of Education* (New York: Free Press, 1973), p. 77.

23. Benito Mussolini, "The Doctrine of Fascism," in *Social and Political Philosophy*, ed. John Somerville and Ronald E. Santoni (Garden City, N.Y.: Anchor Books, 1963), p. 426.

24. Adolf Hitler, *Mein Kampf*, trans. Ralph Mannheim (Boston: Houghton Mifflin, 1971), p. 394.

25. This proclamation was displayed prominently at the Eleanor Roosevelt Centennial Conference, Vassar College, October 1984, as exemplary of both Franklin and Eleanor Roosevelt's ideals.

26. Howard, *Causes of War*, pp. 26–28.

27. William Graham Sumner, "The Conquest of the U.S. by Spain," in *War and Other Essays*, ed. Albert Galloway Keller (Freeport, N.Y.: Essay Index Reprint Series, 1970), pp. 297–334.

28. Sandra Gilbert, "Soldier's Heart: Literary Men, Literary Women, and the Great War," *Signs: Journal of Women in Culture and Society* 8 (3 [1983]):422–50; quotation on p. 433. See, as well, the discussion of the Pankhursts in Anne Wiltscher, *Most Dangerous Women* (London: Pandora Press, 1985), pp. 36–42.

29. Caroline E. Playne, *Society at War, 1914–1916* (Boston: Houghton Mifflin, 1931), pp. 47, 85. Mrs. Humphrey Ward quoted in Playne from p. 17 of *England's Effort*.

30. Quoted in Playne, *Society at War*, p. 119. For an account of militarism in this epoch, see Alfred Vagts, *A History of Militarism: Roman and Realities of a Profession* (New York: W. W. Norton, 1937), pp. 246–316.

31. Playne, *Society at War*, p. 136.

Notes

32. Cited in Playne, *Society at War*, p. 134.

33. Quoted in Ida Husted Harper, *History of Woman Suffrage*, vol. 5 (New York: J. J. Little & Ives, 1922), pp. 578–79.

34. Ibid.

35. David M. Kennedy, *Over Here: The First World War and American Society* (New York: Oxford University Press, 1980), p. 29.

36. Randolf Bourne, "The State," in *The Radical Will: Randolf Bourne: Selected Writings 1911–1918*, ed. Olaf Hansen (New York: Urizen Books, 1977), pp. 354–95; quotations on pp. 17, 145.

37. Cited from Wilson's third annual message to Congress in Kennedy, *Over Here*, p. 24.

38. George Will, "The Real Campaign of 1984," *Newsweek* (2 Sept. 1985):80.

39. Kennedy, *Over Here*, pp. 61, 62.

40. Ibid., p. 68.

41. Cited in Carol S. Gruber, *Mars and Minerva: World War One and the Uses of Higher Learning in America* (Baton Rouge: Louisiana State University Press, 1975), p. 255.

42. Kennedy, *Over Here*, pp. 82, 165–66.

43. H. C. Peterson and Gilbert C. Fite in *Opponents of War, 1917–1918* (Madison: University of Wisconsin Press, 1957), detail the activities of these and other nationalist associations, public, quasi-public, and private. Quotation on p. 46.

44. Kennedy, *Over Here*, pp. 84–86.

45. Ibid., pp. 53ff.

46. Gruber, *Mars and Minerva*, pp. 130–31.

47. Quotation from ibid., p. 96. This book is chock full of the details.

48. Ibid., p. 259.

49. Cited in Kennedy, *Over Here*, p. 246.

50. Randolf Bourne, "The State," p. 361.

Chapter 4. The Attempt to Disarm Civic Virtue

1. Roland H. Bainton, *Christian Attitudes Toward War and Peace: A Historical Survey and Critical Re-evaluation* (New York: Abingdon Press, 1960), p. 53.

2. This perspective appears in, for example, "The UN Address," *Origins* 9 (11 Oct. 1979):258–66; "To Serve Peace, Respect Freedom: World Day of Peace Message," *Origins* 10 (8 Jan. 1981):465–70; "The Cause of Peacemaking in 1983," *Origins* 12 (13 Jan. 1983):490–92; "Recalling the Teaching of Mahatma Gandhi," *Origins* 15 (20 Feb. 1986):586–87.

3. Bainton, *Christian Attitudes*, p. 53.

4. Peter Brock, *Pacifism in Europe to 1914* (Princeton: Princeton University Press, 1972), p. 3.

5. Ibid., pp. 4–5.

6. See René Girard, "Generative Violence and the Extinction of Social Order," *Salmagundi* 63–64 (Spring–Summer 1984):204–37, for a brilliant exegesis of "The Demons of Gerasa."

Notes

7. Peter Brock tells the tale in greater detail in *Pacificism*, p. 13.

8. The text is available in several works, including H. J. Musurillo, *The Acts of Christian Martyrs* (New York: Oxford University Press, 1972). Peter Dronke, in *Women Writers of the Middle Ages: A Critical Study of Texts from Perpetua to Marguerite Porete* (Cambridge: Cambridge University Press, 1984), offers a felicitous commentary attuned to Perpetua's voice.

9. All Augustine quotes are drawn from Augustine, *The City of God*, ed. David Knowles (Baltimore, Md.: Penguin Books, 1972).

10. This reading owes much to Peter Brown's masterful biography, *Augustine of Hippo* (Berkeley: University of California Press, 1969), and his essay, "Political Society," in R. A. Markus, ed., *Augustine: A Collection of Critical Essays* (Garden City, N.Y.: Anchor Books, 1972), pp. 311–29. Quote from Augustine, *City of God*, is from Book XIX, chap. 24, p. 890.

11. See Bainton, *Christian Attitudes*, p. 110.

12. Norton Downs, ed., *Basic Documents in Medieval History* (Princeton: D. Van Nostrand, 1959), pp. 73–76.

13. Christine de Pizan, *The Book of the City of Ladies*, trans. Earl Jeffrey Richards (New York: Persea Books, 1982), pp. 40–42. See also Katharina M. Wilson, *Medieval Women Writers* (Athens: University of Georgia Press, 1984), pp. 333–60.

14. Natalie Zemon Davis, "Men, Women and Violence: Some Reflections on Equality," *Smith Alumnae Quarterly* (April 1977):12–15.

15. Christine de Pizan, *The Book of Three Virtues*, in Katharina M. Wilson, ed., *Medieval Women Writers* (Athens: University of Georgia Press, 1984), pp. 350–53.

16. "Instruction on Christian Freedoms and Liberation," Congregation for the Doctrine of the Faith, *Origins* 14 (17 April 1986):714–28.

17. Cited in Bainton, *Christian Attitudes*, p. 201.

18. Ibid., p. 207.

19. Sarah Gordon, *Hitler, Germans and the 'Jewish Question'* (Princeton: Princeton University Press, 1984), pp. 260–61.

20. Michael Howard, *The Causes of War* (Cambridge, Mass.: Harvard University Press, 1984), p. 27.

21. The Countess of Warwick, *A Woman and the War* (New York: George H. Doran, 1916), pp. 101–13.

22. Quoted in Sandra Gilbert, "Soldier's Heart: Literary Men, Literary Women, and the Great War," *Signs: Journal of Women in Culture and Society* 8 (3 [1983]):422–50; quotation on p. 422.

23. Peter Brock, *Pacifism in the United States: From the Colonial Era to the First World War* (Princeton, N.J.: Princeton University Press, 1968), p. 81.

24. Wilfred Owen, "Anthem for Doomed Youth," in *The Oxford Book of War Poetry*, ed. Jon Stallworthy (New York: Oxford University Press, 1984), p. 188.

25. G. W. F. Hegel, *The Phenomenology of Spirit*, trans. A. V. Miller (Oxford: Clarendon Press, 1977), p. 400.

26. Davis, "Men, Women and Violence," p. 15.

27. George Steiner, *Antigones* (New York: Oxford University Press, 1984), p. 28.

28. Hegel, *Phenomenology*, pp. 274–75. The sphere of the Family and the "immediacy of desire" is the Law governing woman's ethical life, argues Hegel. The universal is reserved to the male who is "sent forth."

Notes

29. Eileen Power, "The Position of Women," in *The Legacy of the Middle Ages*, ed. C. G. Crump and E. F. Jacob (Oxford: Clarendon Press, 1948), pp. 432–33.

30. John M. Todd, *Luther: A Life* (New York: Crossroad, 1982), p. 319.

31. The Countess of Warwick, *A Woman and the War*, pp. 25–26, 54–55, 149, 151–52.

32. Mrs. F. S. Hallowes, *Mothers of Men and Militarism* (London: Headley Brothers, Bishopsgate), pp. 22, 24–25, 47, 107.

33. Barbara Zanotti, "Patriarchy: A State of War," in *Reweaving the Web of Life*, ed. Pam McAllister (Philadelphia: New Society Publishers, 1982), pp. 16–19.

34. Barbara L. Baer, "Apart to the End?" *Commonweal* (22 March 1985):167.

35. Mary Beard, *Woman as Force in History* (New York: Collier Books, 1972), p. 48.

36. Ibid.

37. See Richard J. Niebanck, *Conscience, War and the Selective Objector* (Board of Social Ministry: Lutheran Church of America, 1972).

38. Michael Walzer, *Just and Unjust Wars* (New York: Basic Books, 1977), pp. 256–57.

39. John Paul II, "The UN Address," *Origins* 9 (11 Oct. 1979):266.

40. Hannah Arendt, *The Human Condition* (Chicago: University of Chicago Press, 1958), pp. 238–42.

41. Freeman Dyson, *Weapons and Hope* (New York: Harper & Row, 1984), pp. 4–5.

42. Walzer, *Just and Unjust Wars*, p. xv.

43. Ibid., p. 274.

44. Ibid., p. 264.

45. See, for example, "What Does Church Teach on War, Peace, Nuclear Arms?" *The Pilot* (26 February 1983):8 (*The Pilot* is the paper of the Catholic Archdiocese of Boston). See also Kenneth A. Briggs, "Bishops Taking the Letter on Atomic War to the Parishes," *New York Times*, 16 December 1983, p. B1.

46. Paul Boyers, *The Bomb's Early Light* (New York: Pantheon Books, 1985), pp. 228, 218.

47. Ibid., pp. 183–84.

48. Ibid., pp. 184–85.

49. John W. Dower, *War Without Mercy: Race and Power in the Pacific War* (New York: Pantheon Books, 1986), p. 54, tells the whole sorry tale from both sides.

50. Reinhold Niebuhr, *Christianity and Power Politics* (New York: Archon Books, 1969), p. 23.

51. Robert F. Kennedy, *Thirteen Days* (New York: W. W. Norton, 1969), p. 31.

52. See Gregory S. Kavka, "Morality and Nuclear Politics: Lessons of the Missile Crisis," in Steven Lee and Avner Cohen, eds., *Nuclear Weapons and the Future of Humanity* (Totowa, N.J.: Rowman & Allanheld, 1986), pp. 233–54.

Notes

Chapter 5. Women: The Ferocious Few/
The Noncombatant Many

1. Jean Bethke Elshtain, "On Beautiful Souls, Just Warriors and Feminist Consciousness," *Women's Studies International Forum* 5 (1982):341–48.

2. Hannah Arendt, *The Human Condition* (Chicago: University of Chicago Press, 1958), p. 182.

3. I have summarized the story told by James Axtell, "The Vengeful Women of Marblehead: Robert Roules's Deposition of 1677," *William and Mary Quarterly* 31 (1974):647–52. I am grateful to John Demos for sending me this reference detailing an episode of women's violence.

4. Natalie Zemon Davis, "Men, Women, and Violence: Some Reflections on Equality," *Smithsonian Quarterly* (April 1977):12–15.

5. J. M. Taylor, *Eva Perón: The Myths of a Woman* (Chicago: University of Chicago Press, 1979), p. 16. See also Peggy Reeves Sandy, *Male Dominance, Female Power* (Cambridge: Cambridge University Press, 1981).

6. Davis, "Men, Women, and Violence," p. 14.

7. See Davis's comments on this conundrum in her suggestive essay (ibid.).

8. George Dumézil, *The Destiny of the Warrior*, trans. Alf Hiltebeitel (Chicago: University of Chicago Press, 1969), pp. 106–7.

9. Marguerite Duras, *War* (New York: Pantheon Books, 1986), pp. 23–24.

10. James Reston, Jr., *Sherman's March and Vietnam* (New York: Macmillan, 1984), p. 215.

11. All material from Plutarch is drawn from *Moralia* III, trans. Frank Cole Babbit (Cambridge, Mass.: Harvard University Press, 1983), pp. 473–582.

12. Nancy Huston, "The Matrix of War: Mothers and Heroes," *Poetics Today* 6 (1985):153–70, offers these insights on symmetries in death recognitions for birthing women and killing/killed men (pp. 164–65).

13. Jack Weston, *The Real American Cowboy* (New York: Schocken Books, 1985), pp. 166–84.

14. Elizabeth F. Ellet, *The Women of the American Revolution*, vol. 2 (New York: Haskell House, 1969), pp. 123, 126. Originally published in 1850. The three volumes in this series detail women's patriotic resolve, sacrifices, and general imperturbability in the midst of "heart-chilling fears" and dire suffering. Women are portrayed as loyal Children of the Republic. Most—but not all—are devoted wives and mothers.

15. Ibid., pp. 127, 130, 134.

16. Rita Mae Brown, *High Hearts* (New York: Bantam, 1986).

17. James Axtell, *The Invasion Within* (New York: Oxford University Press, 1985).

18. Information is drawn from a pamphlet published in 1973, underwritten by Jim Beam Distillery, for the state of New Hampshire, entitled "Hannah Duston, Heroine of the 1697 Massacre of Indian Captors on River Islet at Boscawen, N.H."

19. This quote and others in this paragraph are drawn from Shelley Saywell, *Women in War* (New York: Viking Press, 1985), pp. 67, 146.

Notes

20. Duras, *War*, pp. 134–35, 115.

21. Shelley Saywell, *Women in War* (New York: Viking, 1985), p. 38.

22. Ibid.

23. Milovan Djilas, *Wartime*, trans. Michael B. Petrovich (New York: Harcourt, Brace, Jovanovich, 1977), p. 210.

24. K. Jean Cottam, "Soviet Women in Combat in World War II: The Ground Forces and The Navy," *International Journal of Women's Studies* 3 (14 [1980]):345–57; quotation on p. 345. See also Cottam's follow-up piece, "Soviet Women in World War II: The Rear Services, Resistance Behind Enemy Lines and Military Political Workers," *International Journal of Women's Studies* 5 (4 [1980]):363–78).

25. See Richard Holmes, *Acts of War* (New York: Free Press, 1985), p. 102.

26. Saywell, *Women in War*, pp. 149, 132.

27. Ibid., p. 73.

28. See Eric J. Leed's extraordinary *No Man's Land* (Cambridge: Cambridge University Press, 1979) for more on this subject.

29. Mary Beard, *Woman as Force in History* (New York: Collier Books, 1972), p. 288.

30. Evan S. Connell, *Son of the Morning Star* (San Francisco: North Point Press, 1984), pp. 14, 54.

31. Michel Foucault, *Discipline and Punish* (New York: Vintage Books, 1979).

32. Quoted in Saywell, *Women in War*, pp. 215–16.

33. Lynda van Devanter, *Home Before Morning* (New York: Warner Books, 1983), pp. 117, 357.

34. Elizabeth Cady Stanton, Susan B. Anthony, and Matilda Joslyn Gage, eds., *History of Woman Suffrage*, vol. 2 (Rochester, N.Y.: Charles Mann, 1887), pp. 23, 1–2.

35. Mary Elizabeth Massey, *Bonnet Brigades* (New York: Alfred A. Knopf, 1966), p. 367.

36. Ida Husted Harper, ed., *History of Woman Suffrage*, vol. 5 (New York: J. J. Little & Ives, 1922).

37. See Barbara J. Steinson, *American Women's Activism in World War I* (New York: Garland Publishing, 1982).

38. Joseph P. Lash, *Eleanor and Franklin* (New York: New American Library, 1971), p. 281.

39. Ibid., p. 310.

40. See Eleanor Roosevelt, *This Is My Story* (New York: Harper & Bros., 1937), p. 260; and Joseph P. Lash, *Love, Eleanor: Eleanor Roosevelt and Her Friends* (Garden City, N.Y.: Doubleday, 1982), p. 67.

41. Mrs. Franklin D. Roosevellt (*sic*), *It's Up to the Woman* (New York: Frederick A. Hokes, 1933), pp. 243, 263. See also Susan Ware, "ER and Democratic Politics," in Joan Hoff-Wilson and Marjorie Lightman, eds., *Without Precedent: The Life and Career of Eleanor Roosevelt* (Bloomington, Ind.: Indiana University Press, 1984), p. 49; Jean Bethke Elshtain, "Eleanor Roosevelt as Activist and Thinker," *Halcyon 1986* 8 (1986):93–114.

42. *History of Woman Suffrage*, vol. 2, pp. 1–2.

43. This figure, and all others cited, is drawn from D'Ann Campbell's meticulous

analysis in *Women at War with America* (Cambridge, Mass.: Harvard University Press, 1985).

44. Ibid., p. 165.

45. Ibid., p. 174.

46. On how the Nazis fared with women, see Leila J. Rupp, *Mobilizing Women for War* (Princeton, N.J.: Princeton University Press, 1978), and Jill Stephenson, *The Nazi Organization of Women* (Totowa, N.J.: Barnes and Noble, 1981). Quote is in Stephenson, p. 170.

47. *Life Special Issue, World War II* 8 (Spring-Summer 1985).

48. Ibid., p. 86.

49. In Robert Graves, *Good-Bye to All That* (New York: Jonathan Cape & Harrison Smith, 1929), pp. 271–74.

50. Ibid.

51. Ibid., pp. 274–75.

Chapter 6. Men: The Militant Many/
The Pacific Few

1. Tzvetan Todorov, *The Conquest of America* (New York: Harper Colophon, 1984), p. 91.

2. Ibid., p. 92.

3. Mary Beard, *Woman as Force in History* (New York: Collier Books, 1972), pp. 288–89.

4. See the argument in her *Missile Envy* (New York: Bantam Books, 1984).

5. " 'Rambo Style' Dress Upsets Some Vets," *Daily Hampshire Gazette* (Northampton, Mass.), 4 October 1985, p. 15.

6. "Nightclub for the Fatigued Patriot," *Washington Post*, 19 December 1986, p. C11.

7. "Resurgence of Military Toys Brings Profits and Concern," *Daily Hampshire Gazette* (Northampton, Mass.), 17 August 1983, p. 22.

8. Timothy K. Smith, "War Games: In Alabama's Woods, Frank Camper Trains Men to Repel Invaders," *Wall Street Journal*, 19 August 1985, p. 1.

9. Ibid.

10. Poll data and quote from Jean Bethke Elshtain, "The Politics of Gender," *The Progressive*, 22 February 1984, pp. 22–25, on p. 23.

11. See Sigmund Freud, "Thoughts for the Times of War and Death," *The Standard Edition of the Complete Psychological Works of Sigmund Freud*, vol. 14 (London: Hogarth Press, 1964), pp. 275–88; and Sigmund Freud, "Why War?" *Standard Edition*, vol. 20, pp. 199–215.

12. All quotes are drawn from J. Glenn Gray, *The Warriors: Reflections on Men in Battle* (New York: Harper & Row, 1970), pp. 53–54.

13. William Broyles, Jr., "Why Men Love War," *Esquire Magazine*, November 1984, pp. 55–65. Broyles has read Gray.

14. Gray, *The Warriors*, p. 148.

Notes

15. Gordon Zahn, *Another Part of the War: The Camp Simon Story* (Amherst, Mass.: University of Massachusetts Press, 1979), p. 251.

16. Studs Terkel, *The Good War* (New York: Ballantine, 1984), p. 165 (from the testimony of John H. Abbott).

17. Freeman Dyson, *Disturbing the Universe* (New York: Harper Colophon, 1979), pp. 30–32. For an exhaustive history of pacifism in this country, see Peter Brock, *Pacifism in the United States: From the Colonial Era to the First World War* (Princeton, N.J.: Princeton University Press, 1968).

18. Max Hastings, *Bomber Command* (New York: Dial Press, 1979), p. 352.

19. Natalie Zemon Davis, "Men, Women and Violence: Some Reflections on Equality," *Smith Alumnae Quarterly*, April 1977, p. 13.

20. Ernst Jünger, *The Storm of Steel* (New York: Howard Fertig, 1975), p. 316.

21. See Richard Holmes, *Acts of War* (New York: Free Press, 1985), p. 300.

22. S. L. A. Marshall, *Men Against Fire* (New York: William Morrow, 1947), pp. 77–78.

23. See John Keegan, *The Face of Battle* (New York: Penguin Books, 1978).

24. Gray, *The Warriors*, pp. 46–47.

25. Erich Marie Remarque, *All Quiet on the Western Front* (New York: Fawcett, 1975), p. 236.

26. From Siegfried Sassoon's *Memoirs of an Infantry Officer* (New York: Coward McCann, 1931), p. 39.

27. Yoshida Mitsuru, *Requiem for Battleship Yamato*, trans. Richard H. Minear (Seattle: University of Washington Press, 1985), p. 45. See also "War Generation," a splendid short story by Shusaku Endo in his collection *Stained Glass Elegies* (New York: Dodd, Mead, 1985), pp. 147–58.

28. Yukio Mishima, "Patriotism," in *Death in Midsummer* (New York: New Directions, 1966), pp. 93–118.

29. Quotations from "Families Talk of Victims' Hopes, Fears," *Boston Globe*, 30 August 1983, p. 11; and Cable Network News, 30 August 1983.

30. Gray, *The Warriors*, p. 83.

31. Robert Graves, *Good-Bye to All That* (New York: Jonathan Cape & Harrison Smith, 1929), p. 250. See also Holmes, *Acts of War*, pp. 106–8, for further examples.

32. Remarque, *All Quiet*, p. 61.

33. Evan S. Connell, *Son of the Morning Star* (San Francisco: North Point Press, 1984), p. 118.

34. PBS Special, "Return to Iwo Jima," 1985.

35. William Broyles, Jr., *Brothers in Arms* (New York: Alfred A. Knopf, 1986), pp. 273, 275.

36. Vera Brittain, *Testament of Youth* (New York: Wideview Books, 1980), p. 291.

37. Vera Brittain, *Chronicle of Youth: The War Diary, 1913–1917* (New York: William Morrow, 1982), pp. 89–90.

38. Ibid., pp. 101, 308.

39. Stephen Crane, *The Red Badge of Courage* (New York: D. Appleton-Century, 1926; originally published 1894); A. P. Herbert, *The Secret Battle* (London: Hutchinson, 1919); James Jones's trilogy—*From Here to Eternity* (New York: Dell, 1983; originally published 1951), *The Thin Red Line* (New York: Dell, 1985; originally published 1962), *Whistle* (New York: Dell, 1985; originally published 1978); Joseph Heller, *Catch-*

Notes

22 (New York: Simon & Schuster, 1961); Guy Sajer, *The Forgotten Soldier* (New York: Harper & Row, 1967).

40. Paul Fussell, *The Great War and Modern Memory* (New York: Oxford University Press, 1975).

41. Paul Fussell, review of *The Oxford Book of War Poetry*, "The Muse at War," *Boston Globe*, 2 October 1984, p. 14.

42. Jayne Anne Phillips, *Machine Dreams* (New York: Pocket Books, 1984); Bobbie Ann Mason, *In Country* (New York: Harper & Row, 1985).

43. Isabel Colegate, *The Shooting Party* (New York: Avon Books, 1982), pp. 20–21.

44. The story appears in *The Stories of Muriel Spark* (New York: E. P. Dutton, 1985), pp. 263–69.

45. Gertrude Stein, *Wars I Have Seen* (New York: Random House, 1945).

46. Sandra Gilbert, "Soldier's Heart: Literary Men, Literary Women, and the Great War," *Signs: Journal of Women in Culture and Society* 8(3 [1983]:446).

47. Fussell, *Great War*, p. 310.

48. Edith Wharton, *Fighting France: From Dunkerque to Belfort* (New York: Charles Scribner's Sons, 1915), p. 15.

49. Ibid., pp. 53, 54, 220.

50. W. E. B. Griffin, *Brotherhood of War* (New York: Jove Books, 1986); William Young Boyd, *The Gentle Infantryman* (New York: St. Martin's, 1985); Scott C. S. Stone, *Pearl Harbor* (Honolulu: Island Heritage Publications, 1977).

51. All quotes in the following discussion are drawn from Willa Cather, *One of Ours* (New York: Alfred A. Knopf, 1922).

52. Ernest Hemingway, "War Medals for Sale," in *By-line: Ernest Hemingway*, ed. William White (New York: Charles Scribner's Sons, 1967), pp. 120–23; quotation on p. 120.

53. Ernest Hemingway, "Soldier's Home," in *The Short Stories of Ernest Hemingway* (New York: Charles Scribner's Sons, 1953), pp. 143–53; quotation on p. 145.

54. Ernest Hemingway, *A Farewell to Arms* (New York: Charles Scribner's Sons, 1929), pp. 177–78.

55. Broyles, *Brothers in Arms*, p. 137.

56. The Vietnam literature is vast. The best book to come out of the war remains Michael Herr's *Dispatches* (New York: Avon, 1978). Books that consist of collections of remembrances strung together with narrative proliferate, including: Peter Goldman and Tony Fuller, *Charlie Company: What Vietnam Did to Us* (New York: Ballantine Books, 1983); Mark Baker, *Nam: The Vietnam War in the Words of the Soldiers Who Fought There* (New York: Berkley Books, 1981); Wallace Terry, *Bloods: An Oral History of the Vietnam War by Black Veterans* (New York: Ballantine Books, 1984). Personal narratives include: Robert Mason, *Chickenhawk* (New York: Viking, 1983); Rick Eilert, *For Self and Country* (New York: William Morrow, 1983); Ron Kovic, *Born on the Fourth of July* (New York: McGraw-Hill, 1976).

57. Making the war *familiar* is the intent, and the effect, of *Dear America: Letters Home from Vietnam*, ed. Bernard Edelman (New York: W. W. Norton, 1985). Turning the war into therapeutic problems, and solutions, is the genre represented by Robert Jay Lifton's *Home from the War* (New York: Colophon Books, 1973); and Arthur Egendorf's *Healing from the War* (Boston: Houghton Mifflin, 1985).

58. Herr, *Dispatches*, p. 244.

59. Ibid., p. 67.

60. Veterans cited in James Reston, Jr., *Sherman's March and Vietnam* (New York: Macmillan, 1984), p. 183.

61. Hemingway, "Soldier's Home," p. 143.

62. Eric J. Leed, *No Man's Land: Combat and Identity in World War I* (Cambridge: Cambridge University Press, 1979).

63. Gray, *The Warriors*, pp. 171–213.

64. Leed, *No Man's Land*, p. 3.

65. Ibid., pp. 12–13.

66. Remarque, *All Quiet*, p. 13. The physical concreteness of war *is* the texture of Norman Mailer's *The Naked and the Dead* (London: Granada, 1979). Originally published in 1948.

67. Remarque, *All Quiet*, pp. 54, 55.

68. Leed, *No Man's Land*, pp. 35–36.

69. Ibid., p. 8.

70. Ibid., pp. 8, 106.

71. Ibid., p. 163.

72. Remarque, *All Quiet*, p. 143.

Chapter 7. Neither Warriors nor Victims: Men, Women, and Civic Life

1. See especially Albert Camus, *Resistance, Rebellion, and Death* (New York: Alfred A. Knopf, 1961).

2. Freeman Dyson, *Weapons and Hope* (New York: Harper & Row, 1984), pp. 4–5.

3. Michael Howard, *War and the Liberal Conscience* (London: Temple Smith, 1978).

4. Montesquieu, cited in Albert Hirschman, *The Passions and the Interests* (Princeton, N.J.: Princeton University Press, 1978), p. 80.

5. Joseph Schumpeter, *Imperialism and Social Classes*, trans. Heinz Norden (New York: Meridian, 1955).

6. Hirschman, *The Passions and the Interests*.

7. John Stuart Mill, cited in Howard, *War and the Liberal Conscience*, p. 48.

8. Howard, *War and the Liberal Conscience*, p. 53.

9. James's essay is reprinted in William James, *Memories and Studies* (New York: Longmans, Green, 1911), pp. 267–96; quotations in this paragraph from pp. 267, 269, 272, 276, 284, 290.

10. See Anne Wiltsher, *Most Dangerous Women: Feminist Peace Campaigners of the Great War* (London: Pandora, 1985); Jill Liddington, "The Women's Peace Crusade: The History of a Forgotten Campaign," in Dorothy Thomson, ed., *Over Our Dead Bodies: Women Against the Bomb* (London: Virago, 1983), pp. 180–98; Barbara J. Steinson, "The Women's Peace Party: New Departures and Old Arguments," in Dorothy G. McGuigan, ed., *The Role of Women in Conflict and Peace* (Ann Arbor: University of Michigan Press, 1977), pp. 45–53; Linda Schott, "The Women's Peace Party and the Moral Basis for Women's Pacifism," in *Frontiers: A Journal of Women's Studies* 8

Notes

(2 [1985]):18–24; Jane Addams, *Peace and Bread in Time of War* (New York: Macmillan, 1922).

11. Addams, *Peace and Bread*, p. 7.

12. Ibid., p. 129.

13. Virginia Woolf, *Three Guineas* (New York: Harcourt Brace Jovanovich, 1966).

14. Jane Addams, *Newer Ideals of Peace* (London: Macmillan, 1907); idem, *The Long Road of Woman's Memory* (New York: Macmillan, 1916).

15. Addams, *Long Road*, pp. 82, 126.

16. Woolf, *Three Guineas*, p. 113.

17. See Elaine Showalter, *A Literature of Their Own: British Women Novelists from Brontë to Lessing* (Princeton, N.J.: Princeton University Press, 1977), p. 295.

18. Nancy Huston, "The Matrix of War: Mothers and Heroes," *Poetics Today* 6 (1–2 [1985]):153–70. Quote on p. 168.

19. Marilyn French, *The Women's Room*, cited in Huston, "Matrix of War," p. 168.

20. Phyllis Chesler, *Women and Madness*, cited in Huston, "Matrix of War," p. 168.

21. Monique Wittig, *Les Guérillères* (Boston: Beacon Press, 1985), p. 115.

22. Huston, "Matrix of War," p. 168.

23. Mary Daly, cited in Barbara Zanotti, "Patriarchy: A State of War," in Pam McAllister, ed., *Reweaving the Web of Life* (Philadelphia: New Society Publishers, 1982), pp. 16–19.

24. Sally Miller Gearhart, "The Future—If There Is One—Is Female," in McAllister, ed., *Reweaving the Web of Life*, pp. 266–84.

25. Barbara Love and Elizabeth Shanklin, "The Answer is Matriarchy," in Joyce Trebilcot, ed., *Mothering: Essays in Feminist Theory* (Totowa, N.J.: Rowman and Allanheld, 1983), pp. 275–83.

26. The brief is available from the NOW Legal Defense and Educational Fund, 132 West 42nd Street, New York, N.Y. 10036.

27. Cynthia Enloe, *Does Khaki Become YOU? The Militarisation of Women's Lives* (London: Pluto Press, 1983), pp. 16–17.

28. See Sara Ruddick's essays "Maternal Thinking" and "Preservative Love and Military Destruction: Some Reflections on Mothering and Peace," in Trebilcot, ed., *Mothering*, pp. 213–30 and 231–62, respectively.

29. Eric J. Leed, *No Man's Land* (Cambridge: Cambridge University Press, 1979), p. 5.

30. Richard Halloran, "Women, Blacks, Spouses Transforming the Military," *New York Times*, 25 August 1986, p. 1.

31. See Luix Overbea, "Growing Numbers of Black Women See Military Service as Ticket to a Stable Career," *Christian Science Monitor*, 26 September 1985, p. 3; figures from Halloran, "Women, Blacks, Spouses," p. 1.

32. Antonio Handler Chayes, in testimony before the House Armed Services Committee, quoted in Seth Cropsey, "Women in Combat?" *Public Interest* (Fall 1980):58–73.

33. Carolyn Becraft, cited in Halloran, "Women, Blacks, Spouses," p. 1.

34. "The Changing Profession," prepared by Air Force ROTC Recruiting Division (Maxwell Air Force Base, Alabama, 1981), p. 11.

35. See Esther F. Fein, "The Choice: Women Officers Decide to Stay in or Leave," *New York Times Magazine*, 5 May 1985, pp. 32–46.

Notes

36. Betty Friedan, *The Second Stage* (New York: Summit, 1981).

37. A good source is Helen Rogan, *Mixed Company: Women in the Modern Army* (Boston: Beacon Press, 1981).

38. "Women in the Armed Forces," *Newsweek*, 18 February 1980, pp. 34–42.

39. For discussions on the modern citizen-soldier, see Eliot A. Cohen, *Citizens and Soldiers: The Dilemmas of Military Service* (Ithaca: Cornell University Press, 1985); and Morris Janowitz, *The Reconstruction of Patriotism* (Chicago: University of Chicago Press, 1983).

40. Helen Caldicott, *Missile Envy* (New York: Bantam Books, 1985), pp. 1, 316–17.

41. Jonathan Schell, *The Fate of the Earth* (New York: Alfred A. Knopf, 1982).

42. See, for example, Vaclav Havel, "Peace: The View from Prague," *New York Review of Books*, 21 November 1985, pp. 28–30.

43. Adam Smith, *An Inquiry into the Nature and Causes of the Wealth of Nations* (New York: Modern Library, 1985).

44. Adam Ferguson, *An Essay on the History of Civil Society* (Edinburgh: Edinburgh University Press, 1966).

45. Ibid., p. 136.

46. J. G. A. Pocock, *The Machiavellian Moment* (Princeton, N.J.: Princeton University Press, 1975), p. 513.

47. George Kateb, "Nuclear Weapons and Individual Rights," *Dissent* (Spring 1986): 161–72.

48. John Diggins, *The Lost Soul of American Politics* (New York: Basic Books, 1984), p. 315.

49. See ibid., pp. 327–28, for additional insights on these themes; Lincoln cited on pp. 321, 331.

50. Adam Michnik, "Maggots and Angels," in *Letters from Prison and Other Essays*, trans. Maya Latynski (Berkeley: University of California Press, 1986), pp. 169–98.

51. George Orwell, "Notes on Nationalism," in *The Collected Essays, Journalism and Letters of George Orwell. As I Please 1943–1945*, ed. Sonia Orwell and Ian Angus (New York: Harcourt Brace Jovanovich, 1968), pp. 361–80; quotation on p. 363.

52. Adam Clymer, "Polling Americans," *New York Times Magazine*, 10 November 1985, p. 37.

53. Robert Graves, *Good-Bye to All That* (New York: Jonathan Cape & Harrison Smith, 1929), p. 277.

54. Studs Terkel, *The Good War* (New York: Ballantine, 1984), p. 109.

55. J. Glenn Gray, *The Warriors: Reflections on Men in Battle* (New York: Harper Colophon, 1970), p. 217.

56. Leed, *No Man's Land*, p. 4.

57. Quoted in Leed, *No Man's Land*, p. 61.

58. Immanuel Kant, *Perpetual Peace and Other Essays*, trans. Ted Humphrey (Indianapolis: Hackett, 1983); the Kantian future citation is from Joseph H. Nye, Jr., *Nuclear Ethics* (New York: Free Press, 1986); other references in quotes from Carla Pasquinelli, "Is There an Anthropology of Peace and War?" *Cultural Roots of Peace* (Zurich: Gottlieb Duttweiler Institute, World Future Studies Federation) 34 (June 1984):9–20.

59. Sheldon Wolin, "Hannah Arendt: Democracy and the Political," *Salmagundi* (Spring–Summer 1983):3–19; quotation on p. 17. See also Jean Bethke Elshtain, "Re-

Notes

flections on War and Political Discourse: Realism, Just War, and Feminism in a Nuclear Age," *Political Theory* 13 (1 [February 1985]):39–57.

60. Paul Virilio/Sylvere Lotringer, *Pure War*, trans. Mark Polizotti (New York: Columbia University, Semiotexte(e), Inc., 1983).

61. Mark Juergensmeyer, *Fighting with Gandhi* (New York: Harper & Row, 1984), is a useful addition to a massive Gandhi bibliography.

INDEX

278

Index

Index

F-15 jet-fighters, 59–60
Field nursing, 183
Fighting France (Wharton), 214
Fighting Sullivans, The (film), 191*n*
Finch, Francis M., 28
First World War, *xii*, 18, 43, 85, 94, 106–20, 140, 166, 183, 186, 192, 194, 206, 224, 229, 254; British volunteers in, 138; computer simulation of, 89; conscientious objectors in, 204; disillusionment in, 248; Eleanor Roosevelt and, 187–88; French conscription in, 64; Freud on, 200; Lenin on, 82; literature of, 211–12, 214–18; sanctification of war effort in, 137; women's peace efforts in, 233
"First Year of My Life, The" (Spark), 213
Fonda, Jane, 166*n*
Ford, John, 94
Forgotten Soldier, The (Sajer), 222
Foucault, Michel, 92, 182
Four-Minute Men, 115–16
Fowle, Elida Rumsey, 102
Franco-Prussian War, 64, 81
Fredegund, Queen, 135*n*
Frederick III, Emperor, 79
French, Marilyn, 238
French Republic, conscription in, 63–66
French Resistance, 170, 176–77
French Revolution, 62–63, 67, 69, 70*n*, 73, 182; Hegel on, 73–74
Freud, Sigmund, 82*n*, 199–200
Friedan, Betty, 243
Friendly Persuasion (film), 29
From Here to Eternity (film), 30
F-16 jet-fighters, 9
Fussell, Paul, 213
Futurists, 85–86

Gage, Matilda Joslyn, 185
Galard, Geneviève de, 184
Gambrill, J. Montgomery, 118
Gandhi, Mohandas, 23, 28–29, 125*n*, 139, 204
Garner, James, 38
Gaudiam et Spes (Pastoral Constitution on the Church in the Modern World), 149
Geertz, Clifford, 4

Gentle Infantryman, The (Boyd), 216*n*
Georgia, Sherman's march through, 101
Germanic tribes, 181, 196
G.I. Joe (toy), 197
G.I. Joe Action Stars (cereal), 198–99
Gilbert, Sandra, 112
Goldman, Emma, 5*n*
Gold Star Mothers of Future Wars, 41
Gone with the Wind (film), 95–96
Goodbye, Darkness (Manchester), 206
Good-Bye to All That (Graves), 192, 194, 205
Goodman, Ellen, 231
Good War, The (Terkel), 180, 204
Gordon, Mary, 41
Grant: A Biography (McFeely), 195
Graves, Robert, 192–94, 205, 209, 213, 225, 254
Gray, J. Glenn, 10, 191, 196, 200–201, 207, 209, 213, 222, 254
Great War and Modern Memory, The (Fussell), 213
Greece, ancient, 47–56, 166, 230; Rousseau on, 61
Green Berets (film), 12
Greenham Common Women's Peace Camps, 233, 240
Grenada, U.S. invasion of, 199
Griffin, W. E. B., 216*n*
Grotius, Hugo, 87, 150
Group violence, female, 167–71
Gruber, Carol S., 119
Guevara, Che, 85
Guilt, 222; collective, 33*n*
Gulf of Tonkin incident, 36
Gung Ho (magazine), 198*n*
Gun ownership, 23–24

Hafkesbrink, Hannah, 255*n*
Hahne, Dellie, 180
Hale, Nathan, 27–28, 214
Hallowes, Mrs. F. S., 146
Hanson, Donald W., 88*n*
Harris polls, 219*n*
Hartsock, Nancy, 51*n*
Harvard University, 88, 219*n*
Hasbro Industries, 197
Hastings, Max, 204*n*
Hawks, Howard, 43

Index

Hecuba, 98
Heflin, Van, 52n
Hegel, Georg Wilhelm Friedrich, 4, 73–77, 80, 83, 118, 140, 141n, 142
Heigon, Mugg, 167
Hélias, Pierre-Jakez, 63, 65–66
Heller, Joseph, 212
Hemingway, Ernest, 209, 212, 216–18, 221, 222
Heraclitus, 103, 166
Herbert, A. P., 212
Here Is Your War (Pyle), 21
Herr, Michael, 220
Higgins, Marguerite, 20, 22, 181, 185
Hiroshima, 34, 39, 151, 155
Hirschman, Albert, 228
Hispanic cowboys, 174n
History of Woman Suffrage, The (Stanton, Anthony, and Gage), 104n, 185, 186
Hitler, Adolf, 33, 34, 109, 110, 157, 190, 204
Hobbes, Thomas, 31, 74, 84, 87, 88, 128
Holden, William, 94
Holmes, Oliver Wendell, Jr., 117n
Holmes, Richard, 206
Holy War, 134
Home Defense League, 116
Homer, 49–53, 182
Homeric warrior, 47, 48
Horse Soldiers, The (film), 94–95
House of Commons, 113
Howard, Michael, 63, 77, 90, 111, 228
Howe, Julia Ward, 106
Humanism, 227–28
Hungary, 1956 Russian invasion of, 29, 30
Hunting, 23–24
Husban, Mary Morris, 102
Huston, Nancy, 48, 49, 238

Idealism, 87, 88
Idylls of the King (Tennyson), 97
Iliad (Homer), 49–50
In Country (Mason), 213
Indians, 167–69, 175–76, 191, 196; women's roles among, 181–82
Individualism, 109
Inquiry into the Nature and Causes of the Wealth of Nations, An (Smith), 249

International Farm Youth Exchange, 28
International Journal on World Peace, 229
International relations, 86–91
Iroquois, 196
Israel, 9; 1956 attack on Egypt by, 29, 30
Israeli Defense Forces, 242n
Italian Resistance, 176
Iwo Jima, 210
IWW, 116

James, William, 230–31
Jennings, Peter, 179
Jerusalem Bible, The, 121
Jesus, 122–25, 153n
Jews in U.S. armed forces, 72n
Joan of Arc, 173, 174, 177
Joan of Arc (film), 16
John Paul II, Pope, 123, 152
Johnson, Lyndon B., 36, 37
John XXIII, Pope, 34, 35
Jones, James, 212
Judgment at Nuremberg (film), 33
Jünger, Ernst, 194, 206, 213
Junkers, 76
Just and Unjust Wars (Walzer), 149, 154
Just Warriors, xii, xiii, 4, 6–9, 137–40, 164, 165; Beautiful Souls and, 141n, 142, 148; Christianity and, 47; in movies, 43
Just-war tradition, 149–59

Kamikaze, 206
Kant, Immanuel, 85n, 87, 228n, 255
Kapuscinski, Ryszard, 256
Kateb, George, 250n
Keegan, John, 8, 207
Kellogg-Briand Pact (1928), 234
Kennan, George, 87
Kennedy, David M., 115n, 116, 117
Kennedy, John F., 32, 159; assassination of, 36
Kennedy, Robert F., 37, 38n, 159
King, Martin Luther, Jr., 32, 38, 252
King Edward, the Kaiser and the War (Legge), 112
King Philip's War, 167
Kingston, Maxine Hong, 163

Index

Kiowas, 182
Kirby, Mrs. William, 99
Knights, Christian, 133
Knights of Liberty, 116
Korda, Zoltan, 201
Korean War, 18–20, 171
Kovic, Ron, 219

Labor force, women in, 7
Ladd, Alan, 52n
Lasch, Christopher, 6
Lebanon, 9; Marines killed in, 208
Leed, Eric, 222, 223, 224n, 240, 254, 270n28
Legge, Edward, 112
Lenin, V. I., 81–86
Lennon, John, 37
Letter to M. D'Alembert on the Theatre (Rousseau), 67
Levée en masse, 63, 67, 182
Leviathan (Hobbes), 88
Liberalism, 227–30
Liberalism and the Limits of Justice (Sandel), 38n
Liberty League, 116
Life of Coriolanus (Plutarch), 56
Life magazine, 20, 37, 190, 191
Lilia, 135n
Lincoln, Abraham, 18–19, 97, 106, 108, 250–51
London *Morning Post*, 192
Longest Day, The (film), 35
Long Road of Woman's Memory, The (Addams), 235
Losey, Carol, 208
Losey, Second Lieutenant Donald, 208
Louis XVI, King of France, 62–63
Love in Germany, A (film), 68n
Lovelace, Richard, 97
Luther, Martin, 135–36, 142–43, 149
Lynchings, 116, 170
Lysistrata (Aristophanes), 50

MacArthur, General Douglas, 181n
Macaulay, Rose, 112
McDonnell Douglas, 59
McFeely, William S., 195

McGuire, Mrs., 98
Machiavelli, Niccolò, 31, 33, 35–36, 48, 49, 54n, 56–61, 70, 87, 128, 131, 149, 251
Machiavellianism (Meinecke), 87
Machiavellian Moment, The (Pocock), 57n
Machine Dreams (Phillips), 213
MAD ("mutual assured destruction"), 49
"Maggots and Angels" (Michnik), 251n
Mailer, Norman, 23n, 212, 274n66
Manchester, William, 206
Manicheanism, 3
"Many Sisters to Many Brothers" (Macaulay), 112
Mao Zedong, 81, 84–85
Marianne, 70
Marines, U.S., 177–78, 206–7; killed in Lebanon, 208
Mars, 10, 172
Marshall, S. L. A., 207
Martyrdom, Christian, 126–27; of Joan of Arc, 173
Marx, Karl, 74n, 80–84
Mason, Bobbie Ann, 213
Massey, Mary Elizabeth, 185
Masu, Marisa, 176, 179
"Maternal thinking," 240
Maximilianus, 125
Mazzini, Joseph, 109, 110
Meinecke, Friedrich, 87
Mein Kampf (Hitler), 110
Men in Dark Times (Arendt), 245
Mercenaries, 198
Mexican War, 81
Michnik, Adam, 251n
Militant Many, 171, 195–202
Mill, John Stuart, 228
Milton, John, 253
Minerva, 172
Minerva (journal), 189n
Mishima, Yukio, 208n
Mitsuru, Yoshida, 208
Mobilizing Women for War (Rupp), 7
Money, Sex and Power (Hartsock), 51n
Montesquieu, 228
Moore, Frank, 103
Moralia (Plutarch), 62, 172
Moral Man and Immoral Society (Niebuhr), 87
More, Thomas, 227
Morgenthau, Hans, 87

283

Index

Mothers of Men and Militarism (Hallowes), 146

Mothers' Strike for Peace, 221

Movies: images of war in, 30; impact on popular culture of, 12; see also specific titles

Murphy, Audie, 12

Mussolini, Benito, 109

Nagasaki, 34, 151, 155

Naked and the Dead, The (Mailer), 274n66

Napoleon, 76

National American Woman Suffrage Association (NAWSA), 113–14, 119, 186

National Board for Historical Service, 117–18

National Committee on the Cause and Cure of War, 234

National Education Association, 117

National Guard, 17

National Industrial Conference Board, 117

Nationalism, 107–8; American, 113–20, 252; and dream of unity, 256; patriotism versus, 97; popular, 108–13

National Organization for Women (NOW), 232, 239

National Patriotic Relief Society, 186

National Security League, 116, 186; Committee on Patriotism through Education of, 117

Nation-state: and civic identity, 73–80; Protestantism and, 135–38, 143

Navy, U.S., 105, 191n; Civil Engineer Corps of, 241

Nazism, 7, 33–34, 110; just-war theory and, 155; movie images of, 30; state church and, 137; women's response to, 190n

New Breed (magazine), 198n

Newer Ideals of Peace (Addams), 235

"New patriotism," 251–52

Newsweek, 243

New Testament, 122, 123, 127n; peace churches and, 139

New York Call, 117

New York Herald Tribune, 20n

New York Times, 10, 12, 116, 244n, 252n

Nicaea, Council of, 125

Niebuhr, Reinhold, 87, 157

Nightingale, Florence, 183

"1945" (Kapuscinski), 256

Nixon, Richard, 37, 39

No Man's Land (Leed), 222

Noncombatant Many, 171, 180–91

Norris, George W., 115n

Nuclear test-ban treaty, 32

Nuclear weapons, 244n; apocalyptic view of, 247; Catholic opposition to, 152–54; just war and, 151; psychological discourse on, 246

Nursing, wartime, 183–84

"Ode to the Morning of Christ's Nativity" (Milton), 253

Officer and a Gentleman, An (film), 11–12

Oglala Sioux, 181, 182

Old Testament, 122, 124, 127

One of Ours (Cather), 216

On Violence (Arendt), 81

On War (Clausewitz), 78

Oresteia (Aeschylus), 50

Origins of Totalitarianism, The (Arendt), 110n

Orwell, George, 217, 252

Over Here (Kennedy), 115

Owen, Wilfred, 140n, 212

Pacem in Terris (papal encyclical), 34, 35n

Pacific Few, 171, 195, 202–10

Pacifism, 139–40; and Catholic just-war tradition, 152; Christian, 122–28; feminist, 8; moral absolutism of, 157; politics and, 248, 249

Paine, Thomas, 227

Pankhurst, Christabel, 112n, 138

Pankhurst, Emmeline, 112

Pankhurst, Sylvia, 112n

Paramilitary training camps, 198

"Patriarchy: A State of War" (Zanotti), 146

Patriotism, 109n; chastened, 252–53, 258; nationalism versus, 97; "new," 251–52

"Patriotism" (Mishima), 208n

Paul, Alice, 148

284

Index

Index

Sandel, Michael, 38n
Sartre, Jean-Paul, 85
Sassoon, Siegfried, 208, 212, 254
Satyagraha, 32
Saxonhouse, Arlene W., 50
Saywell, Shelley, 178
Scandinavian mythology, 172n
Schell, Jonathan, 247
Schenck v. United States, 117n
Schools, during First World War, 117
Schumpeter, Joseph, 228, 229
Scientific Man versus Power Politics (Morgenthau), 87
SDI (strategic defense initiative), 9
Second International, 78
Second Stage, The (Friedan), 243
Second World War, 7, 17–18, 43, 105, 106, 110, 148, 170–71, 206, 254; battle failure in, 207; bombing of German cities in, 155; conscientious objectors in, 202–4; draft in, 138; guilt about, 33–34; homefront in, 189–91; images of enemy in, 201; as just war, 155, 157–58; literature of, 213; mothers in, 41; nationalism in, 109; Pyle on, 20–22; Resistance in, 170, 176–79; Soviet women in combat in, 178
Sedition Slammers, 116
Selective Service System, *see* Draft
Sergeant York (film), 43, 179
Seventh Day Adventists, 204
Shame (film), 38
Shane (film), 52–53
Shaw, George Bernard, 112
Sherman, General William T., 101
Sherman's March and Vietnam (Reston), 171n
Shirer, William L., 33
Shooting Party, The (Colegate), 213
Showalter, Elaine, 236
Sinclair, May, 112
Smeal, Eleanor, 199n
Smith, Adam, 249
Social Contract, The (Rousseau), 61, 69
Socialists, 117
Socrates, 53, 54
Sokokis, 167
Soldier of Fortune (magazine), 198
"Soldier's Home" (Hemingway), 217, 221, 222
Somme, Battle of the, 107

Son of the Morning Star (Connell), 181
Sophocles, 50
"Sources of Soviet Conduct, The" (Kennan), 87
Soviet Union, *see* Russia
Spanish-American War, 111n
Spark, Muriel, 213–14
Sparta: Aristotle on, 54; familialization of army of, 48; honoring of dead in, 173
Spartacists, 84
Spartan mothers, *xiii*, 121, 191, 193; Civil War women as, 99–101, 103, 104; Rousseau on, 62, 69–71, 93
Speech, freedom of, 117
Spellman, Cardinal, 42
Springsteen, Bruce, 219
Stalinism, 110
Stanton, Elizabeth Cady, 5, 6, 104n, 185, 188–89, 191
"Star Warriors," 198
"Star Wars" antiballistic-missile defense proposal, 9
"State, The" (Bourne), 119
Stein, Gertrude, 14, 211, 214, 215
Steiner, George, 70n, 141n
Stevens, George, 52n
Stone, Scott C. S., 216n
Storm of Steel, The (Junger), 194
Story of My Experiments with Truth (Gandhi), 28
Students' Army Training Corps, 118
Suarez, Francisco, 150
Subhuman image of enemy, 201
Suffragette, The (newspaper), 111, 138
Suffragists, 5–6, 144, 145, 188, 233–34
Sullivan, Aletta, 191
Sullivan, Betsy, 98–99
Sumner, William Graham, 111n
Supreme Court, U.S., 72n, 117
Systems dominance, 88

Tacitus, 181, 196
"Ten Commandments of Womanhood, The," 146, 147
Tennyson, Alfred, Lord, 97
Terkel, Studs, 180, 204, 254
Terrible Threateners, 116
Terrorists, 177; women, 178–79
Thatcher, Margaret, 166n

286

Index

Thomas, Augustus, 114–15
"Thoughts for the Times on War and Death" (Freud), 199
Three Guineas (Woolf), 235, 236
Thucydides, 48, 87, 88, 149
Time magazine, 30
Tito, Joseph Broz, 177n
Tocqueville, Alexis de, 56–57, 249
Todorov, Tzvetan, 196
Tokyo Imperial University, 208
"To Locasta, on Going to the Wars" (Tennyson), 97
Tolstoy, Leo, 212
Top Gun (film), 197–98
Toronto Star Weekly, 217
Totalitarian image of enemy, 201
Toys, militaristic, 197
Transnational identifications, 210n
Tree of Heaven, The (Sinclair), 112
Trench warfare, 107
Trojan Women (Euripedes), 50, 51
Truman, Harry S., 19, 34, 38–39
Tuan, Ngo Ngoc, 210

United Daughters of the Confederacy, 100n
United Nations, 34; Security Council, 29
University Peace Congress, 229
Urban II, Pope, 134
USO, 180

Valkyrie, 172n
Van Devanter, Linda, 184
Vatican II, 149
Viet Minh, 184
Vietnam War, 7, 30, 38–39, 110, 166, 200, 210; conscientious objectors in, 202; and Diem, 35; disillusionment in, 248; escalation of, 36; impact of, 9; literature of, 213, 218–21, 273nn56, 57; Medal of Honor recipients in, 207; movies about, 12; nurses in, 183; popular culture images of, 197; war correspondents in, 185; women in, 10
Vigilantism, 116
Vindication of the Rights of Women, A (Wollstonecraft), 69

Virilio, Paul, 32n, 257
Vitoria, Francisco de, 150
Vonnegut, Kurt, 6n

Wajda, Andrzej, 68n
Walker, Lieutenant General Walton H., 180
Waltz, Kenneth, 87n
Walzer, Michael, 149, 152, 154–55, 158
War (Duras), 175, 226
War and Peace (Tolstoy), 212
War correspondents, 185
Ward, Mrs. Humphrey, 112
War Department, U.S., 114n, 118
War in Korea (Higgins), 181
Warriors, The (Gray), 191, 196, 200
Wars I Have Seen (Stein), 14, 211, 214
War toys, 197
Warwick, Countess of, 138, 145–46
Washington, George, 21, 175
Washington Post, 116, 157, 197n, 231
Waskow, Captain Henry T., 20
Waterloo, Battle of, 76
Wayne, John, 12, 94–95, 184, 197, 219
Weapons and Hope (Dyson), 226
Weapons technology, 8–9
Weatherpersons, 84, 178
Weber, Eugene, 63, 64
Weil, Simone, 49
Weintraub, Jeff, 57n
West Germany, women in armed forces in, 242–43
Westmoreland, General William, 242
Weston, Jack, 174n
Wharton, Edith, 214–15
Whiteman Air Force Base, 244n
Whitman, Walt, 113
Why Are We in Vietnam? (Mailer), 23
"Why War?" (Freud), 199–200
Williams, Esther, 16
Wilson, Edmund, 107
Wilson, Woodrow, 112n, 114, 115, 119
Wings, Mary, 11n
Winter Soldier Investigation, 221
Wittig, Monique, 238
Wollstonecraft, Mary, 69–73
Woman and the War, A (Countess of Warwick), 145
Woman as Force in History (Beard), 196

Index